THIS IS REDSTONE MISSILE WEAPON SYSTEM

CHRYSLER CORPORATION MISSILE DIVISION

This text has been digitally watermarked to prevent illegal duplication.

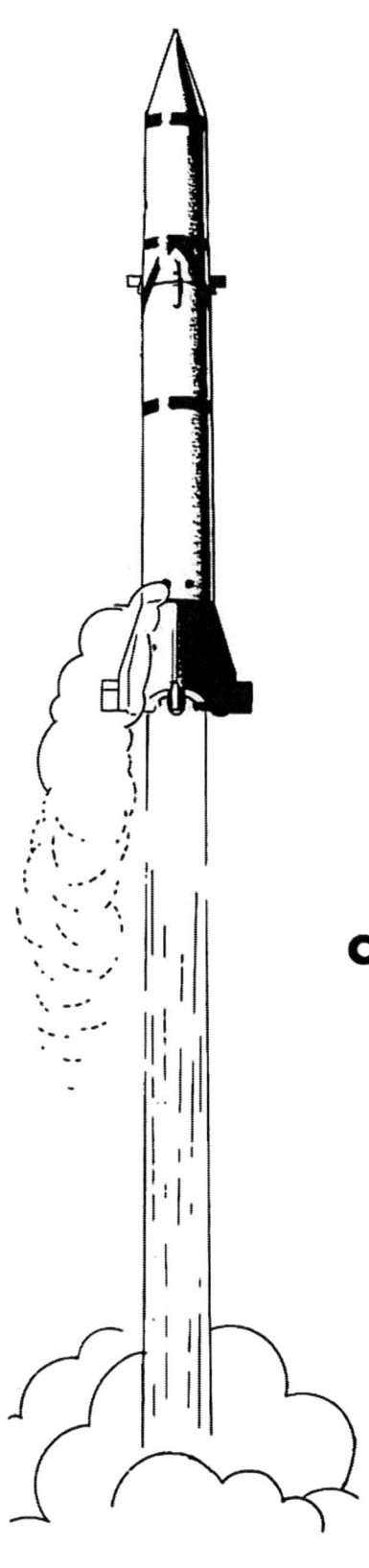

THIS IS REDSTONE

**CHRYSLER CORPORATION
MISSILE DIVISION**

©2012 PERISCOPE FILM LLC
ALL RIGHTS RESERVED
ISBN #978-1-937684-80-8
WWW.PERISCOPEFILM.COM

Frontispiece – Ballistic Guided Missile XM8; Ballistic Shell

TABLE OF CONTENTS

                                                                                  Page

CHAPTER I – REDSTONE HISTORY

GENERAL	I-1
Missile and Space Research	I-1
CHRYSLER'S ROLE	I-2
Making and Assembling the REDSTONE Missile	I-3
TYPE	I-3
FUNCTION AND CAPABILITIES	I-8

CHAPTER II – MISSILE STRUCTURE

GENERAL	II-1
NOSE UNIT	II-3
AFT UNIT	II-4
CENTER UNIT	II-6
TAIL UNIT	II-8
ENGINE MOUNT	II-10

CHAPTER III – GROUND SUPPORT EQUIPMENT

GENERAL	III-1
WARHEAD UNIT SEMITRAILER VAN (XM481)	III-4
AFT UNIT TRAILER (XM480)	III-6
THRUST UNIT TRAILER (XM482)	III-9
ACCESSORIES TRANSPORTATION TRUCK	III-11
ERECTOR-SERVICER (XM478)	III-11
MOBILE PLATFORM LAUNCHER (XM74)	III-12
GUIDED MISSILE PROGRAMER TEST STATION (AN/MSM-38)	III-13
GENERATOR SET	III-15
POWER DISTRIBUTION STATION (AN/MSQ-32)	III-17
BATTERY SERVICING SHOP (XM479)	III-19
AIR COMPRESSOR TRUCK	III-20
AIR SERVICER (XM483)	III-21

TABLE OF CONTENTS (Continued)

Page

CHAPTER III – GROUND SUPPORT EQUIPMENT
(Continued)

ALCOHOL TANK SEMITRAILER (XM388)	III-23
AIR SUPPLY SEMITRAILER	III-24
AIR SEPARATION SEMITRAILER	III-25
LOX TANK SEMITRAILER	III-25
LOX STORAGE CONTAINER	III-26
HYDROGEN PEROXIDE SERVICER (XM387)	III-26
BULK MATERIAL REPAIR PARTS TRUCK (XM486)	III-26
REPAIR PARTS TRUCK (XM488)	III-26
REPAIR PARTS TRAILER (XM487)	III-28
PRESERVATION AND PACKAGING SHOP (XM485)	III-29
PNEUMATIC SHOP (XM477)	III-30
SUPPLY OFFICE (XM484)	III-31
STABILIZED PLATFORM TEST STATION (AN/MJM-2)	III-32
GUIDANCE AND CONTROL COMPONENTS TEST TRAILER "A"	III-34
GUIDANCE AND CONTROL COMPONENTS TEST TRAILER "B"	III-37
CALIBRATION SET – MISSILE SYSTEM TEST EQUIPMENT	III-38
FIRE-FIGHTING EQUIPMENT SET	III-38
GUIDED MISSILE TRAILER, ANALYZER VAN	III-38
GUIDED MISSILE TRAINER (AN/MSQ-T2)	III-40
ST-80 CONTAINER	III-41
LIQUID NITROGEN CONTAINER	III-42
MISSILE LAYING KIT	III-42

CHAPTER IV – GUIDANCE AND CONTROL

GENERAL	IV-1
PROGRAM DEVICE	IV-3
STABILIZATION SYSTEM	IV-5
DESCRIPTION OF COMPONENTS	IV-6
FUNCTION OF THE STABILIZATION SYSTEM	IV-9
GUIDANCE SYSTEM	IV-21
Lateral Guidance	IV-21
Range Guidance	IV-22
Cutoff Computer	IV-23
CONTROL SYSTEM	IV-24
POWER	IV-25
Phase I	IV-25
Phase II	IV-32

TABLE OF CONTENTS (Continued)

Page

CHAPTER V – PROPULSION SYSTEM

GENERAL	V-1
STEAM SYSTEM	V-1
PROPELLANT SYSTEM	V-4
TURBOPUMP	V-5
PNEUMATIC SYSTEM	V-6
THRUST CONTROL	V-6
STARTING SYSTEM	V-7
OPERATIONAL SEQUENCE	V-7
CUTOFF	V-8

CHAPTER VI – PROPELLANT SYSTEMS

GENERAL	VI-1
OXIDIZER SYSTEM	VI-4
Characteristics and Properties of Oxygen	VI-4
FUEL SYSTEM	VI-7
Fuel Transfer System	VI-8
Tank Heater System	VI-9
Inert Fluid Transfer System	VI-10
HYDROGEN PEROXIDE SYSTEM	VI-12

CHAPTER VII – PNEUMATIC SYSTEM

CHAPTER VIII – FIRING SITE OPERATIONS

GENERAL	VIII-1
SITE SELECTION AND PREPARATION	VIII-2
Accessibility	VIII-2
Size of Area	VIII-2
Contour of Terrain	VIII-3
Bearing Strength of Soil	VIII-3
Drainage	VIII-3
Cover and Protection	VIII-3
Survey Control	VIII-3
Lines of Sight	VIII-4
Standby Area	VIII-4
LAYING AND AIMING	VIII-4
PRELIMINARY CONSIDERATIONS AND PREPARATIONS	VIII-5

TABLE OF CONTENTS (Continued)

Page

CHAPTER VIII — FIRING SITE OPERATIONS
(Continued)

GIVEN INFORMATION	VIII-5
EQUIPMENT UTILIZED	VIII-5
PRELIMINARY LAYING	VIII-6
WEAPON SYSTEM EQUIPMENT REQUIRED AT THE FIRING SITE	VIII-7
LINE OF VEHICLE MARCH	VIII-7
EQUIPMENT EMPLACEMENT	VIII-8
HORIZONTAL CHECKOUT	VIII-10
Preparation	VIII-10
Tests Performed	VIII-10
VERTICAL CHECKOUT	VIII-12
Preparation	VIII-12
Tests Performed	VIII-12
FINAL LAYING	VIII-13
Determination of the Magnetic Azimuth of the TLOF	VIII-13
Determination of the ACI Sighting Magnetic Azimuth	VIII-13
Determination of the First ACI Ground Mark	VIII-13
Determination of the Distance Between the ACI and Launcher	VIII-14
Qualification of the ACI Azimuth Scale	VIII-14
Missile Course Alinement	VIII-14
Determination of the Reposition Constant	VIII-14
Repositioning of the ACI	VIII-15
Missile Fine Alinement	VIII-15
FINAL PREPARATIONS FOR FIRING	VIII-15
Preparation for Propellant Loading	VIII-15
Inert Lead Start Fluid Filling	VIII-15
Alcohol Filling	VIII-16
LOX Filling	VIII-16
Hydrogen Peroxide Filling	VIII-16
Vertical Range Computer Test and Presetting	VIII-16
Vertical Lateral Computer Test	VIII-17
Warhead Prelaunch Check	VIII-17
Final System Preparations and Equipment Removal	VIII-17
MISSILE FIRING	VIII-17
RETESTING, ABORT FIRING, AND POST-FIRING OPERATIONS	VIII-19
Retesting	VIII-19
Abort Firing	VIII-19
Post-Firing Operations	VIII-20

TABLE OF CONTENTS (Continued)

Page

CHAPTER IX – TELEMETRY SYSTEM

GENERAL	IX-1
TELEMETRY STANDARDS	IX-1
TYPICAL TRANSMITTER OPERATION	IX-1
TELEMETRIC DATA TRANSMITTING SYSTEM AN/DKT-8 (XO-2)	IX-2
MEASUREMENT	IX-3

LIST OF ILLUSTRATIONS

Figure		Page
Frontispiece	Ballistic Guided Missile XM8; Ballistic Shell	i
I-1	Missile Flight Phases	I-8
II-1	Missile Structure	II-0
II-2	Missile Dimensions	II-1
II-3	Ballistic Missile Shell (Exploded View)	II-2
II-4	Nose Unit	II-3
II-5	Aft Unit	II-5
II-6	Center Unit	II-7
II-7	Tail Unit	II-8
II-8	Multiple Pneumatic Coupling Balcony	II-9
II-9	LOX Replenishing Coupling Balcony	II-9
II-10	Engine Mount and Engine	II-10
III-1	Mobile Launch Site	III-0
III-2	Loading for Shipment by Air	III-2
III-3	Field Deployment	III-3
III-4	Warhead Unit Semitrailer (Left Side)	III-4
III-5	Base Assembly Warhead Unit Trailer	III-5
III-6	Aft Unit Trailer	III-6
III-7	Cover – Aft Unit Trailer	III-7
III-8	Base Assembly – Aft Unit Trailer	III-8
III-9	Thrust Unit Trailer	III-9
III-10	Accessories Transportation Truck (Load A)	III-10
III-11	Accessories Transportation Truck (Load B)	III-10
III-12	Erector-Servicer, Truck Mounted	III-11
III-13	Mobile Launcher and Accessories	III-12
III-14	Guided Missile Programer Test Station (Exterior)	III-13

LIST OF ILLUSTRATIONS (Continued)

Figure		Page
III-15	Guided Missile Programer Test Station (Interior)	III-14
III-16	Generator Set	III-15
III-17	Generator Console Panel	III-15
III-18	Generator Motor Compartment	III-16
III-19	Power Distribution Trailer (Left Rear View)	III-17
III-20	Power Distribution Trailer — Trailer Body (Right Front View)	III-18
III-21	Battery Servicing Trailer	III-19
III-22	Air Compressor Truck	III-20
III-23	Air Servicer	III-21, III-22
III-24	Alcohol Tank Semitrailer	III-23
III-25	Pumping Compartment	III-24
III-26	Air Supply Semitrailer	III-24
III-27	Air Separation Semitrailer	III-25
III-28	LOX Semitrailer	III-25
III-29	Hydrogen Peroxide Servicer	III-27
III-30	Repair Parts Trailer	III-28
III-31	Basic Truck (Bulk Material Repair Parts, Repair Parts, Preservation Packaging)	III-29
III-32	Pneumatic Shop Truck	III-30
III-33	Supply Office and Prime Mover	III-31
III-34	Stabilized Platform Test Station (Exterior)	III-32
III-35	Stabilized Platform Test Station (Interior)	III-33
III-36	Test Trailer A	III-35
III-37	Test Trailer B	III-36
III-38	Guidance and Control Test Trailer (A or B) and Prime Mover (M52)	III-37
III-39	Calibration Set — Missile System Test Equipment	III-38
III-40	Analyzer Van	III-39
III-41	Guided Missile Trainer	III-41
III-42	ST-80 Container	III-41
III-43	Missile Laying	III-43
IV-1	Standard Trajectory and Reference Coordinates	IV-0
IV-2	Guidance and Control Block Diagram	IV-4
IV-3	REDSTONE 6700 Program Device	IV-5
IV-4	Gyroscopic Spin and Precession	IV-10
IV-5	Air Bearing Principle	IV-11
IV-6	Stability of the Platform Ring	IV-12
IV-7	Platform Coordinates	IV-13
IV-8	Alignment Loops	IV-15

LIST OF ILLUSTRATIONS (Continued)

Figure		Page
IV-9	Stabilization Loops	IV-16
IV-10	Accelerometer Loops	IV-19
IV-11	AB-9 Gyro	IV-20
IV-12	Primary Power Source	IV-26
IV-13	A-C Power Distribution	IV-27
IV-14	60-KW Generator	IV-28
IV-15	28-Volt Energizer	IV-29
IV-16	60-Volt Regulated Power Supply	IV-30
IV-17	Mod O Inverter Block Diagram	IV-31
V-1	A-7 Rocket Engine	V-0
V-2	Propellant and Hydrogen Peroxide Flow Diagram	V-2
V-3	Pneumatic System	V-3
V-4	Operation of the Turbopump	V-5
V-5	Thrust Controller	V-7
VI-1	LOX Filling	VI-7
VI-2	Alcohol Fueling	VI-9
VI-3	Inert Fluid System	VI-11
VI-4	Fuel Transfer System	VI-11
VI-5	Drum Heating System	VI-14
VI-6	Hydrogen Peroxide Servicing of the Missile	VI-15
VI-7	Operation of the Steam Generator	VI-16
VII-1	Function of the Air Servicer	VII-2
VII-2	Pneumatic System	VII-3
VII-3	Thrust Unit Pneumatic System	VII-4
VII-4	Body Unit Pneumatic System	VII-5
VII-5	Ground Control System	VII-8
IX-1	Transmitter - Simplified Block Diagram	IX-2

LIST OF TABLES

Table		Page
I-I	Missile Systems	I-4
I-II	REDSTONE and JUPITER C Highlights	I-7
VI-I	Specific Impulse of Some Typical Chemical Propellants	VI-3
IX-I	Oscillator Deviation Chart	IX-4
IX-II	Measuring Program (Typical Missile)	IX-5
IX-III	Measuring Program – Channel 15 Inputs (Typical Missile)	IX-6

This page has been left blank intentionally.

CHAPTER I
REDSTONE HISTORY

This page has been left blank intentionally.

CHAPTER I
REDSTONE HISTORY

GENERAL

During the 1945-50 period of limited funds and limited sense of urgency, the Army built the facilities, assembled the talent, and accumulated the basic knowledge needed to produce its missile systems. At the White Sands Missile Range (then Proving Ground), established in 1944, American and former German missilemen fired V-2's and other missiles under experimental conditions. By 1950 they had fired, in upper-atmosphere experiments, some 67 missiles using V-2 components.

Missile and Space Research

Experimentation performed during those years points up the inseparability of missile and space research. The German V-2, an operational weapon, was used frequently as a space research vehicle. Army's CORPORAL developed in opposite fashion. It began life as a test vehicle and was converted to an operational missile in the early 1950's. In the Fall of 1945, the Army fired its first WAC CORPORAL to an altitude of 43 miles. In February 1946, it launched a BUMPER. This was another test vehicle with a V-2 as first stage and a WAC CORPORAL as the second stage. The BUMPER provided valuable information concerning the separation and ignition of a rocket's second stage in highly rarefied air. It also gave data on the stability of the second stage at extremely high velocities and altitudes and the aerodynamic effects of high Mach numbers. The altitude it attained, 250 miles, stood as a world record for years.

The interrelation of missilery and space activities is personified, of course, by such Army scientists as Drs. von Braun, Stuhlinger, and their colleagues at the Army Ballistic Missile Agency (ABMA). During the 1920's and 1930's these men became interested in rockets and missiles solely as a means of exploring outer space. But the knowledge of rocketry they gained was adapted to military use, first by the German Army, then by the U.S. Army. When these space enthusiasts helped build the JUPITER C, which was used for space exploration, they completed a cycle in their professional lives.

Thus, when the Space Age arrived, a missile-minded Army had the facilities, equipment, and talent to participate in it.

CHRYSLER'S ROLE

During the Korean War, the Navy selected Chrysler to construct and operate a plant to produce Pratt & Whitney J-43 Turbo-Wasp jet engines. Plant construction began on 27 October 1951, and in mid-July, 1952, just before production was to begin, the Navy cancelled the contract. Chrysler completed the 2,100,000 square foot plant, and the Navy assigned this facility to Chrysler for other defense work.

In the summer of 1952, ABMA was seeking a prime contractor for the REDSTONE Missile. Teams were sent to talk to the managements of a number of corporations who had the potential to become prime contractors. In October, 1952, Chrysler received a contract from the Department of Defense to assist ABMA with the design and production of REDSTONE Missiles. The Jet Engine plant was converted into a missile manufacturing facility. Chrysler engineers were integrated into important segments of the Redstone Arsenal at Huntsville, Alabama. This effort was expanded in 1956, when ABMA was activated.

The first REDSTONE Missile was built by the Army and fired in August, 1953, approximately two years after the first studies were initiated.

In November, 1955, the first Chrysler-built missile (designated as number 13) was delivered. Missiles 1 through 12 were built by the Army.

Missile 13 was transported to Huntsville and disassembled to evaluate the production quality. The missile was reassembled and launched a year later (in July of 1956).

The REDSTONE, which was the first ballistic missile to be put into production by Chrysler, is commonly referred to as "the father of American ballistic missiles" and has an unequalled record of successful firings. The 40th Field Artillery Group was the first tactical field unit to be equipped with the REDSTONE. On 16 May 1958, at Cape Canaveral, Florida, Battery A of the 40th Field Artillery Group conducted the first successful troop launching of a REDSTONE Missile.

Making and Assembling the REDSTONE Missile

Mass production concepts stemming from automotive experience have been employed by Chrysler Corporation in the development of fabrication and assembly operations for the REDSTONE Missile. The Chrysler-operated Michigan Ordnance Missile Plant is the only one of its kind operated by a motor car manufacturer and Chrysler is said to be the first U.S. missile builder to place large ballistic missiles in scheduled production.

Chrysler has made enormous strides in the development of facilities, methods, and tooling in this plant. It is a highly organized facility complete with equipment for manufacturing, testing, quality control, and all the elements required to produce a missile ready for deployment to the armed forces here and overseas.

Moreover, a team called the Advanced Projects Organization was formed within the Chrysler Defense Group to specialize in the concept and planning of new weapon and space system projects.

The laboratory testing techniques that were employed in developing the REDSTONE tactical nose cone provide an example of how automotive experience can be harnessed in other directions. Not only was the high-speed re-entry problem solved, but the project was developed on a highly compressed time schedule at low cost. The nose cone was found to be completely successful in its first test firing.

One of the unusual features of this facility is the 60-foot steel tower used for testing missiles under simulated dynamic loading and pressure conditions encountered in actual firings. During testing each missile is filled with a suitable liquid to simulate the density of the LOX and fuel. Fuel tanks are also hydrostatically pressurized to simulate the pressures encountered in flight.

There are many other unique testing devices. Among the more important are: a vertical test stand in the laboratory which moves like a pendulum to simulate the conditions of pitch experienced in flight due to wind force (instrumentation is provided to check on the behavior of control stabilizers which correct for wind force effects during flight); a vibrating "shake" table that duplicates the vibrations experienced during launching; a centrifugal test machine, mounted in a pit, designed to impress loading up to 100g on components; a cone-shaped furnace lined with infrared lamps to study the effect of high temperature on nose cones during reentry; an electronic system for determining the center of gravity of each missile, by components and for the assembled missile.

TYPE

The REDSTONE Missile is an Army Field Artillery Tactical Missile. As a weapon, it is considered to be a long-range surface-to-surface ballistic (projectile) type rocket. As a missile, it is considered to be a medium-range vehicle because it has a range of less than 500 miles.

The thrust necessary to lift the missile off the launcher and to propel the missile during the phase of flight is supplied by means of a single bipropellant liquid rocket (air-independent) engine. This engine develops 78,000 pounds of thrust and propels the missile at supersonic speeds (Mach 4.8). The missile is directed in flight from liftoff to impact by an inertial guidance and control system.

TABLE I-I – Missile Systems

Type and/or Designation		Description and/or Explanation	REDSTONE
Range	Short	Less than 500 miles	x
	Long	Greater than 500 miles (ICBM) and (IRBM)	
Mach No.	Subsonic	Speed less than Mach 1	
	Transonic	Speeds less than Mach 1 to speeds greater than Mach 1	
	Sonic	Mach 1 (speed of sound)	
	Supersonic	Speeds greater than Mach 1	x
	Hypersonic	Speeds greater than relaxation time	
Propulsion	Air independent	Does not use air – Rocket Engine	x
	Air dependent	Uses Air – Jet Engine	
Guidance	Inertial	An automatic navigation system which uses gyroscopic devices.	x
	Preset	Control equipment contained wholly within the missile.	
	Command	Missile's guidance comes from outside the missile. Receiver in missile receives directions from a ground station or mother aircraft.	
	Homing	Usually at end of flight. Radar, heat, and light are homing devices.	

TABLE I-I – Missile Systems (Cont'd)

Type and/or Designation		Description and/or Explanation	REDSTONE
Guidance (Cont'd)	Beam Rider	Missile contains equipment which enables it to follow an electronic beam.	
	Radio Navigation	Consists of master and slave stations that emit low-frequency pulses at constant intervals.	
	Terrestrial	A system of map matching is commonly used. Other systems may use earth gravitational, magnetic, and electrical fields.	
	Celestial	Complex system. A mechanism takes celestial fixes and keeps missile on course electronically.	
Aerodynamic	Projectile	Similar to a bullet	x
	Winged	Airfoil surfaces - similar to airplanes.	
Structure	Reinforced shell	Skin reinforced with complete framework of members.	
	Semimonocoque	Skin reinforced with longerons and bulkheads.	x
	Full monocoque	Skin reinforced with bulkheads.	
Branch of Service	A	Air Force	
	N	Navy	
	G	Army	x
	ANG	All three	
Type	TM	Tactical Missile	x
	SM	Strategic Missile	

TABLE I-I — Missile Systems (Cont'd)

Type and/or Designation		Description and/or Explanation	REDSTONE
Type (Cont'd)	IM	Interceptor Missile	
	GAR	Guided Aircraft Rocket	
	GAM	Guided Aircraft Missile	
Use	SAM	Surface-to-Air Missile	
	AAM	Air-to-Air Missile	
	ASM	Air-to-Surface Missile	
	SSM	Surface-to-Surface Missile	x
	AUM	Air-to-Underwater Missile	
	SUM	Surface-to-Underwater Missile	
	USM	Underwater-to-Surface Missile	
	UAM	Underwater-to-Air Missile	
Status	X	Experimental	
	Y	Service Test	x
	Z	Obsolete	
Test	TV	Test Vehicle	
	A	Aerodynamic	
	C	Control	
	L	Launch	
	P	Propulsion	
	R	Research	
Model	1	First Model	
	2	Second Model	
	3	Third Model	

TABLE I-I – Missile Systems (Cont'd)

Type and/or Designation		Description and/or Explanation	REDSTONE
Modification	a	First Modification	
	b	Second Modification	
	c	Third Modification	

Example: The Regulus 1 Missile – SSM-N-8a
SSM – Surface-to-Surface Missile
N – Navy Buaer
8 – Eighth Model
a – First Modification

TABLE I-II – REDSTONE and JUPITER-C Highlights

Missile	Event	Date Fired
1	First REDSTONE Fired	20 August 1953
13	First Chrysler-built Missile (delivered 14 November 1955)	19 July 1956
27	First Deep Penetration of Space (JUPITER C)	20 September 1956
32	First Chrysler Missile Shipped Directly to AMR	14 March 1957
40	First Nose Cone Recovery (JUPITER C)	8 August 1957
42	First Tactical Top	10 December 1957
29	EXPLORER I (JUPITER C)	31 January 1958
1002	First Troop Firing	16 May 1958
44	EXPLORER IV (JUPITER C)	26 July 1958
50	HARDTACK	1 August 1958
51		12 August 1958

FUNCTION AND CAPABILITIES

Primarily, the REDSTONE Missile is for tactical field operation by U.S. Army personnel, (Field Artillery Missile Battalion) in order to provide general support to a field army, to fulfill the requirements for a medium-range missile, to supplement or extend the range of firepower of existing artillery weapons, to provide increased heavy power fire support for deployed ground combat forces, and to compensate for expanding dimensions of the battle area.

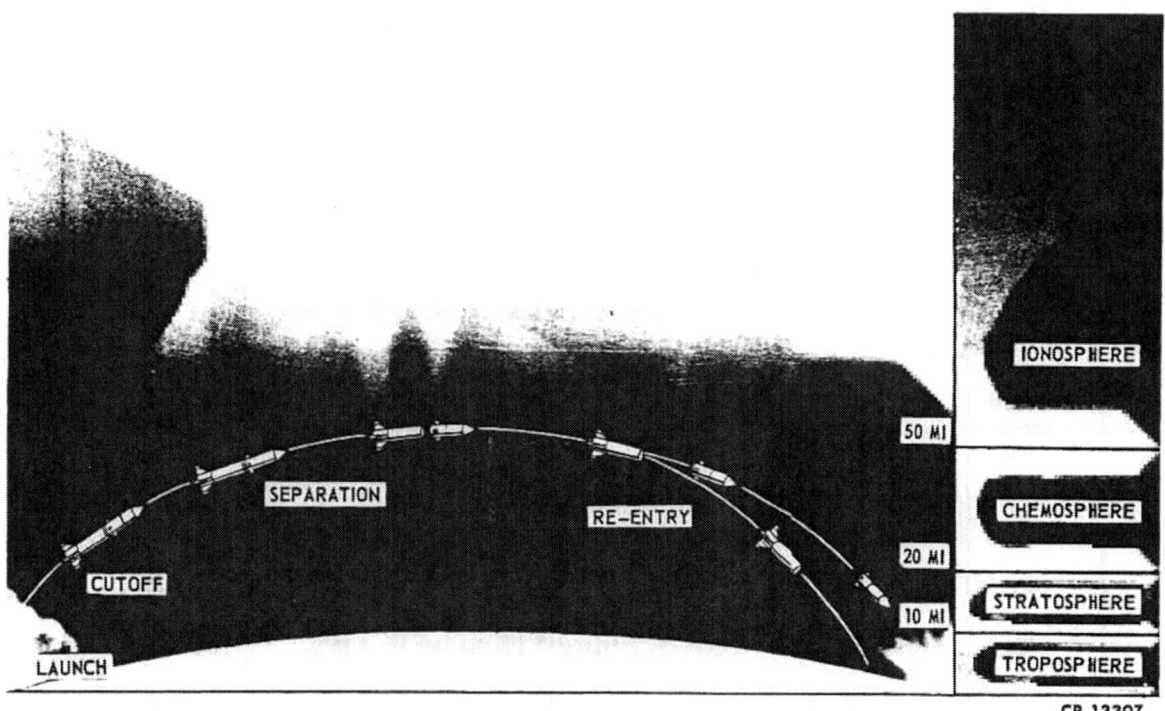

Figure I-1 – Missile Flight Phases

I-8

CHAPTER II
MISSILE STRUCTURE

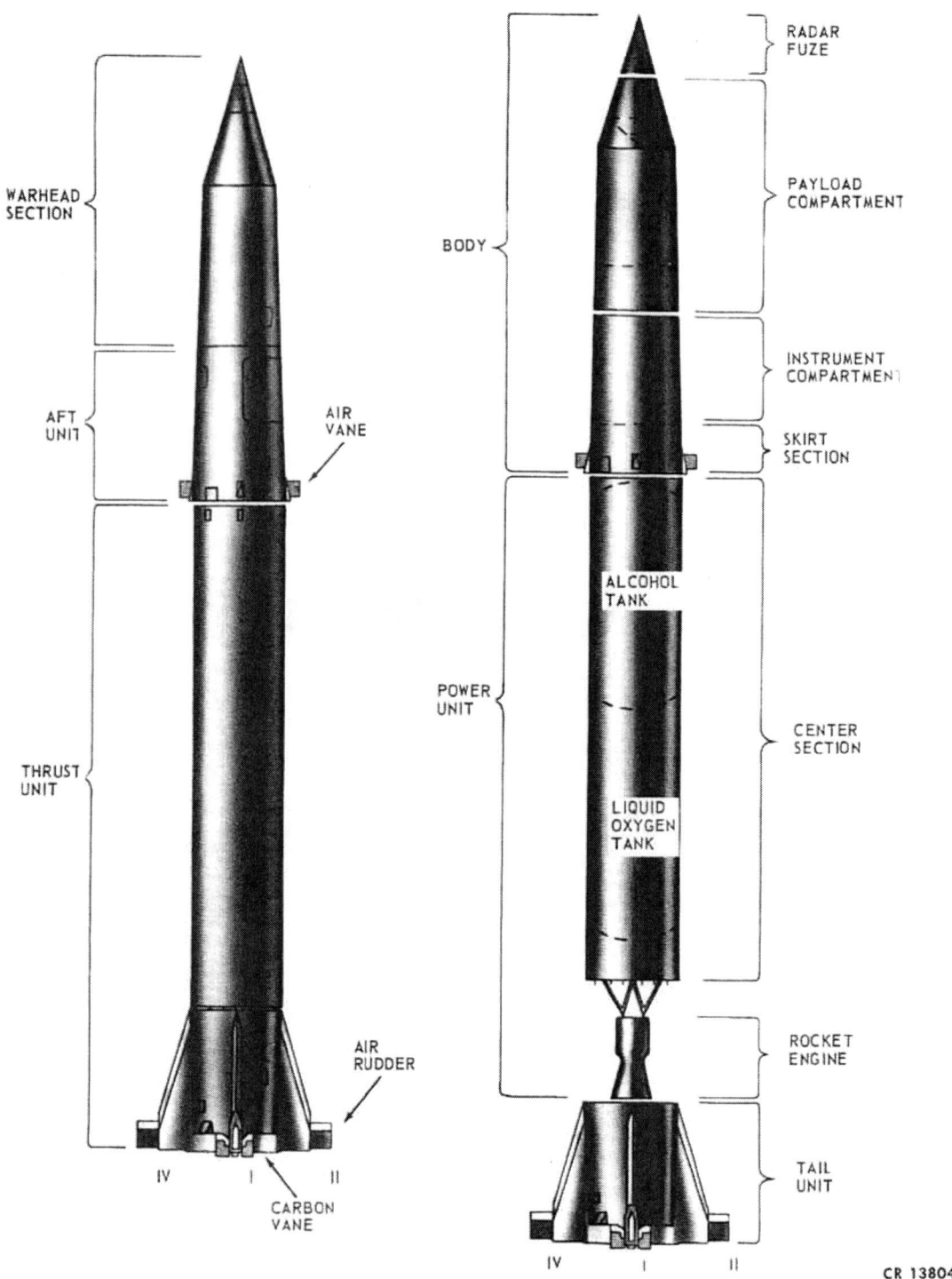

Figure II-1 — Missile Structure

CHAPTER II
MISSILE STRUCTURE

GENERAL

The missile consists of two major parts: the body unit and the thrust unit. The body is subdivided into two main units, the nose (or warhead) unit and the aft unit. The instrument compartment, which contains the guidance and control equipment, is located in the aft unit. The thrust unit is also subdivided into two main units, the center unit and the tail unit, which contain the propellant (LOX and alcohol) tanks and the rocket engine, respectively.

The locations of missile components are referred to as "forward" for components located toward the nose of the missile and "aft" for those located toward the tail. Viewing the missile from the tail, the location of components may also be specified with respect to the air vanes or rudders.

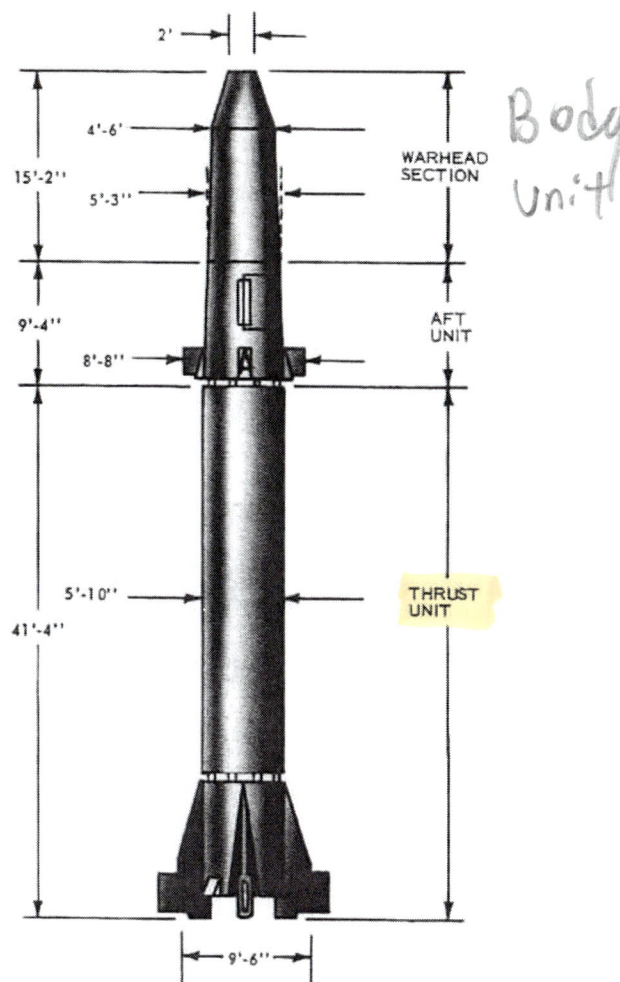

Figure II-2 — Missile Dimensions

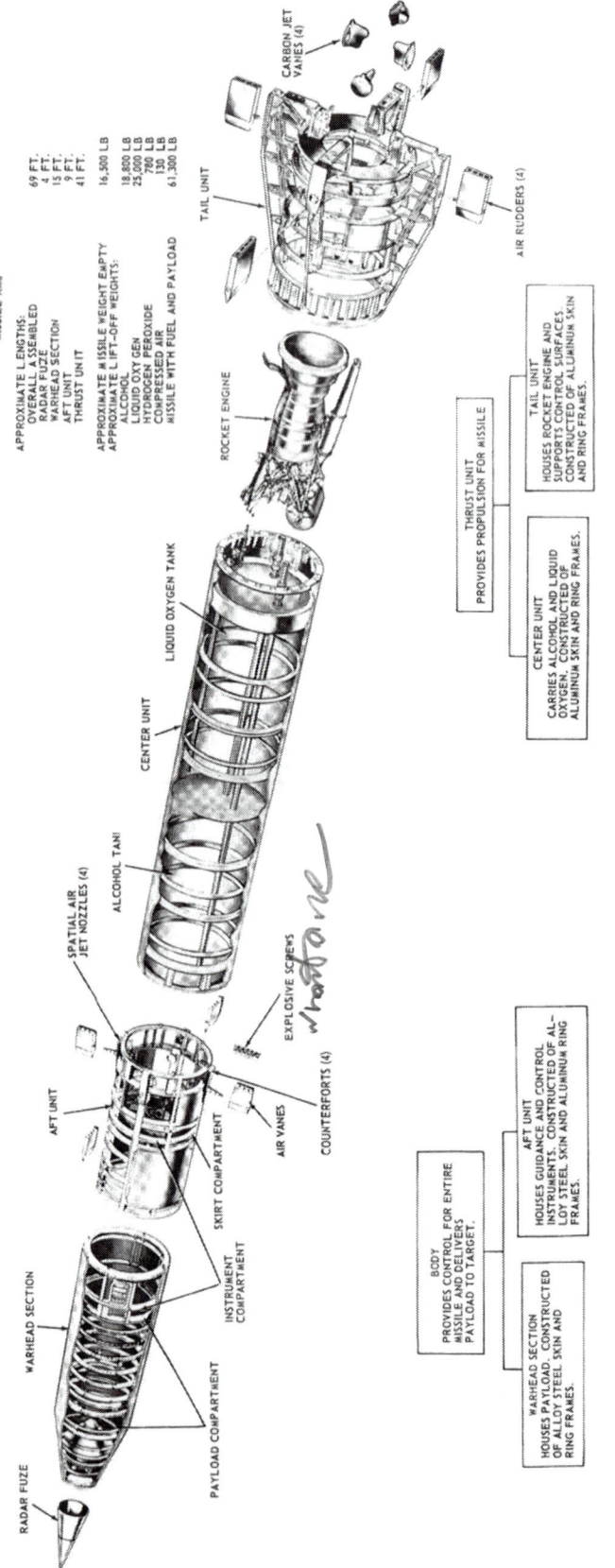

Figure II-3 — Ballistic Missile Shell (Exploded View)

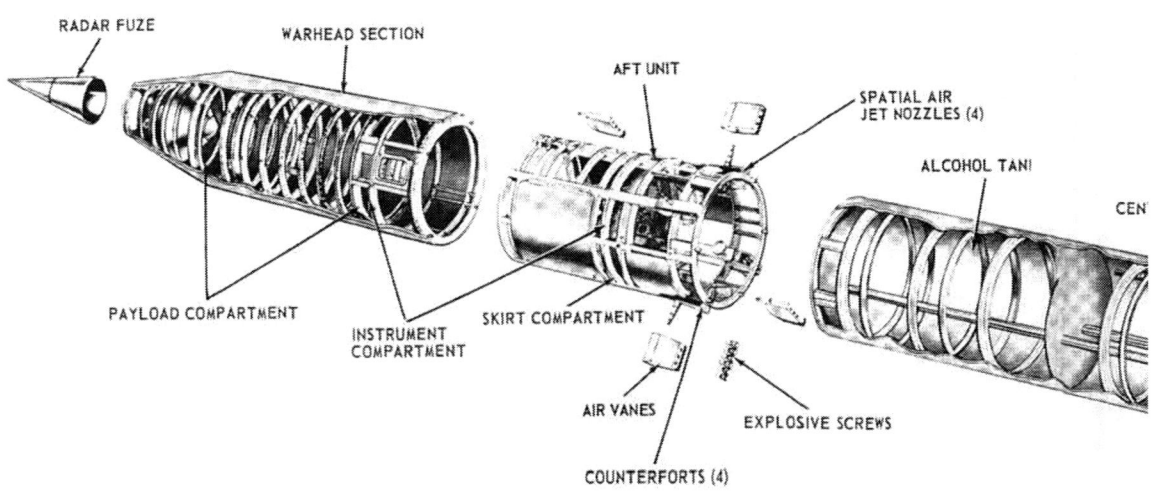

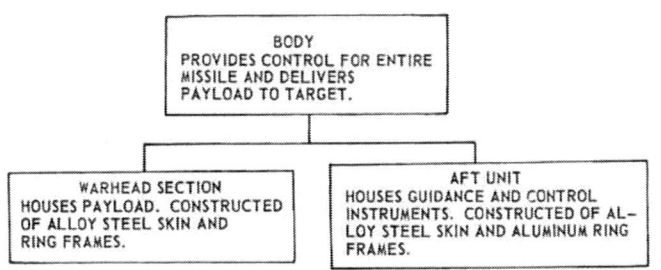

Figure II-3 — Ballistic Missile Shell (Exploded View)

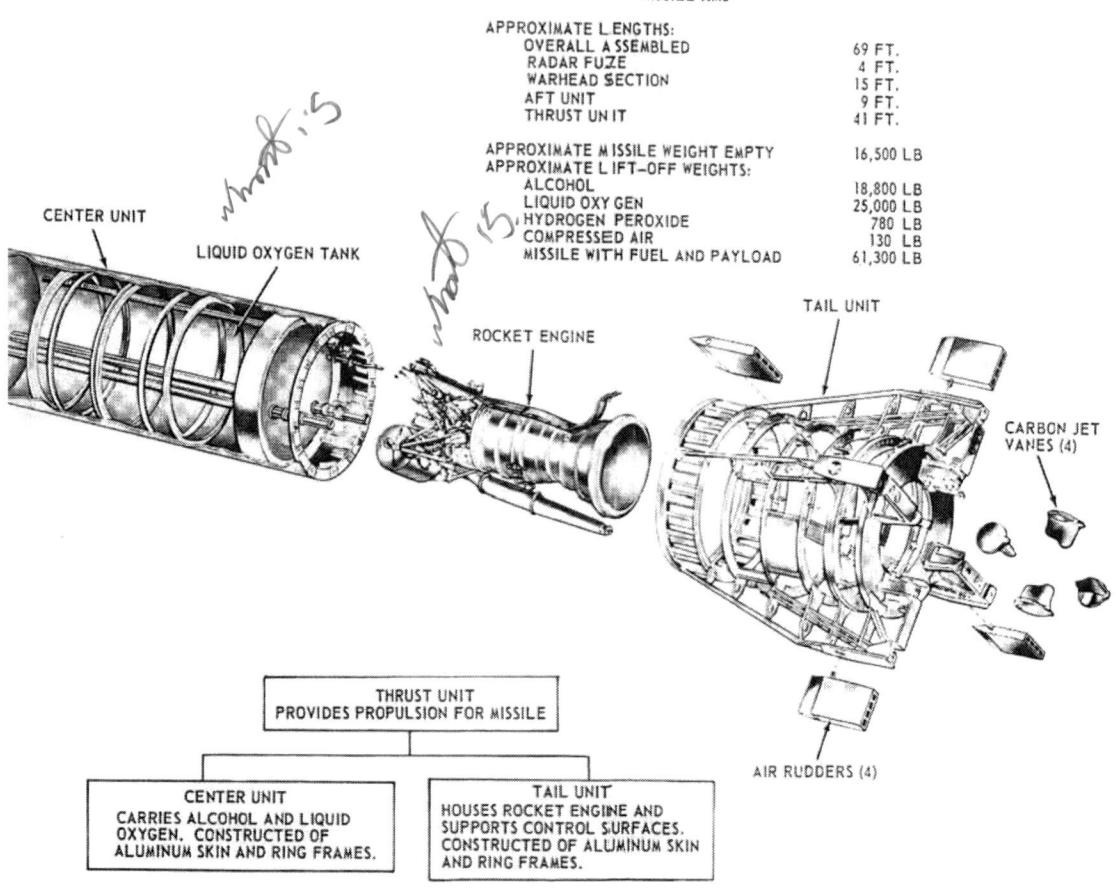

NOSE UNIT

The nose (or warhead) unit consists of a skin of alloy steel that is riveted and welded to a framework of former rings, bulkheads, and stringers. This construction is required in order to enable the nose unit to withstand the high pressure, temperature, and corrosive conditions encountered at re-entry. The pressure reaches 95 psi and the temperature rises to 1000°F on the forward portion of the body (nose unit) when it re-enters the atmosphere. All exterior surfaces are painted with zinc chromate to provide corrosion protection.

The first ring frame is designed as a flange mounting for the fuze cone which is fastened to the nose unit with six machine screws. The last ring of the nose unit has a flanged surface for mating the nose to the aft unit.

A conically shaped metal support, used as one of the mounting surfaces for the payload, is located near the forward end of the nose unit. The aft payload support of the payload compartment is flat, and has a circular opening in the center section. The open center section of this support permits access to the payload compartment for maintenance and is covered by the base of the payload. The payload base is seated against a silicone rubber gasket to provide an airtight seal. This seal is required because the payload support also forms the forward end of the pressurized instrument compartment.

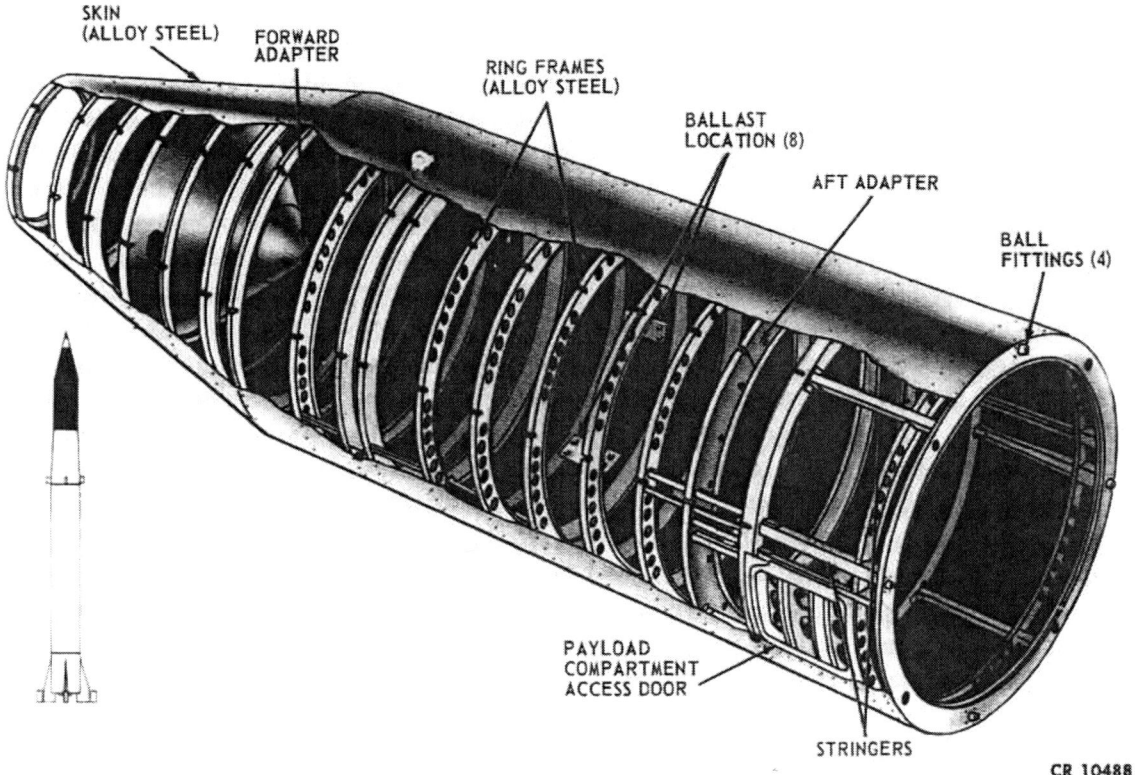

Figure II-4 — Nose Unit

II-3

There are four ball fittings on this former ring that locate and align the nose to the aft unit. These fittings accept the attaching bolts that pass through the first ring frame of the aft to hold the assemblies together. Four additional attaching bolts that do not seat in ball fittings are used at this point. A silicone rubber gasket is fitted between the nose and aft unit to assure airtightness.

Two doors located in the aft end of the nose unit provide access to the payload compartment aft bulkhead. These doors are also fitted with silicone rubber gaskets to assure an airtight fit.

AFT UNIT

The aft unit is a barrel-shaped structure that consists of an alloy steel skin riveted to a framework of <u>aluminum ring frames</u> and stringers. A reinforced pressure bulkhead divides the assembly into two sections: the instrument compartment and the skirt section. The instrument compartment is that portion of the aft assembly that is forward of the pressure bulkhead used to separate the guidance and control compartment from the skirt section. The skirt section is that open-ended portion of the assembly that extends from the pressure bulkhead to the last ring frame. The skirt section houses the two high-pressure air spheres that serve the body pneumatic systems. Four actuators which turn the air vanes are located in the aft end of the skirt section.

Two large doors provide access to the instrument compartment. These doors are fitted with silicone rubber gaskets to assure an airtight seal. The pressure bulkhead forms the aft end of the instrument compartment. An access door is located in the skirt section to provide access to equipment when the body and thrust unit are mated.

Six ball and socket fittings are used to align the body unit with the center unit. The ball fittings are located on the aft former ring of the aft unit, and meet six sockets riveted to the forward former ring of the center unit. Explosive screws are placed in tapped holes in the ball fittings to hold the assemblies together. A similar assembly, using four ball and socket fittings without the explosive screws, aligns the aft unit with the nose unit. At this junction a silicone rubber ring is added to make an airtight seal.

Four counterforts are located on the outer circumference of the skirt section. These counterforts serve as bases for the air vanes and housings for the spatial air jet nozzles. Counterforts II and IV also serve as striking plates for the expulsion cylinders.

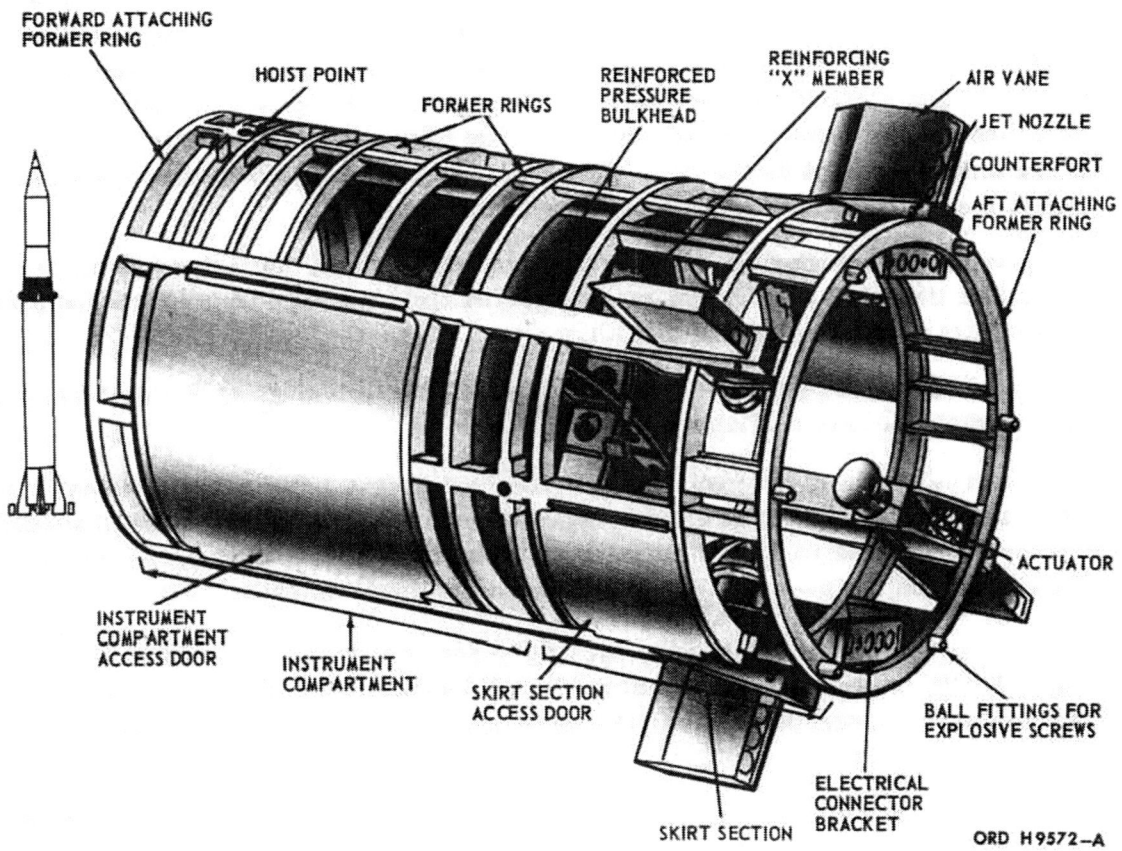

Figure II-5 — Aft Unit

II-5

CENTER UNIT

The propellants are carried in tanks that form the center unit. The center unit is fabricated of aluminum because it does not have to resist the effects of re-entry; this unit will have completed its mission prior to re-entry. The alcohol (alc) and liquid oxygen (LOX) tanks are designed to be pressurized to prevent cavitation and collapse as the propellants are used. Cavitation tends to occur when a high-speed pump tries to move fluids faster than they can flow. When cavitation occurs, a void space of low-pressure air exists between the inlet and outlet of the pump. Were cavitation to occur in the turbopump used on the missile, the pump could not perform the important function of metering the alc and LOX in correct proportions.

A conduit runs through both tanks to carry the general network cable harness and pneumatic lines that connect the pneumatic and electrical components located forward of the tanks with those located in the tail.

The aft bulkhead of the center unit is insulated with fiberglass that is held in place with a dummy bulkhead of aluminum.

The forward attaching former ring of the center unit has six sockets that receive ball fittings on the body to align and fasten the body to the thrust unit. Explosive screw access doors are located aft of the first ring frame and in line with each socket to allow the installation of the explosive screws from outside the missile.

The catwalk is fastened to the first ring frame and extends into the skirt section of the body. It is designed to protect the forward bulkhead and the equipment located on it while work is being performed in the skirt section after the missile has been erected on the launcher.

Four vane lock rod assemblies extend forward from the center section to join the four air vane lock rods in the aft section. These two parts are connected when the center and aft units of the missile are connected. The vane lock rods in the aft unit hold the air vanes in a fixed position until separation of the missile in flight. At separation, the vane lock rods are pulled from the four actuators in the aft unit and the air vanes are left free to operate.

The aft ring frame incorporates four steel pads to which the engine mount is bolted.

Twenty connecting brackets on the aft ring frame fasten the center unit to the tail unit.

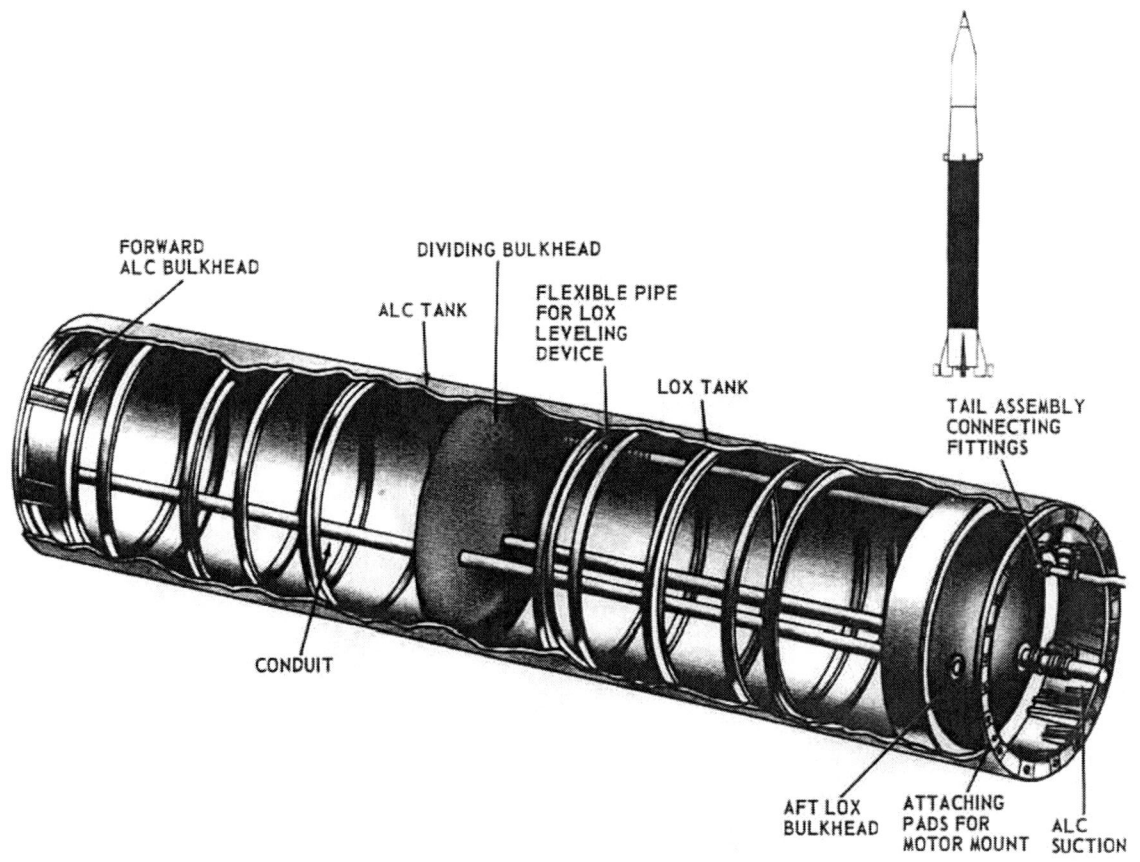

Figure II-6 — Center Unit

TAIL UNIT

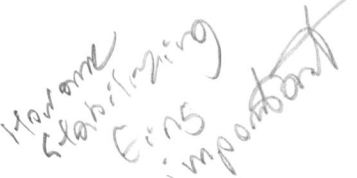

The tail unit is a riveted aluminum structure that consists of four stabilizing fins mounted on a barrel-shaped housing. It is bolted to the aft former ring of the center unit after the propulsion unit has been fastened in place. The tail unit serves to shroud the propulsion unit and also provides control surfaces for the first phases of flight. In addition, the tail unit houses the air-storage spheres and other pneumatic and electrical components. Two access doors are located in the sides of the tail unit.

The high-pressure air spheres that serve the pneumatic system of the thrust unit are mounted in the tail.

Wiring harnesses are mounted on the side of the tail unit and connect the electrical system to the tail plugs located in fins II, III, and IV.

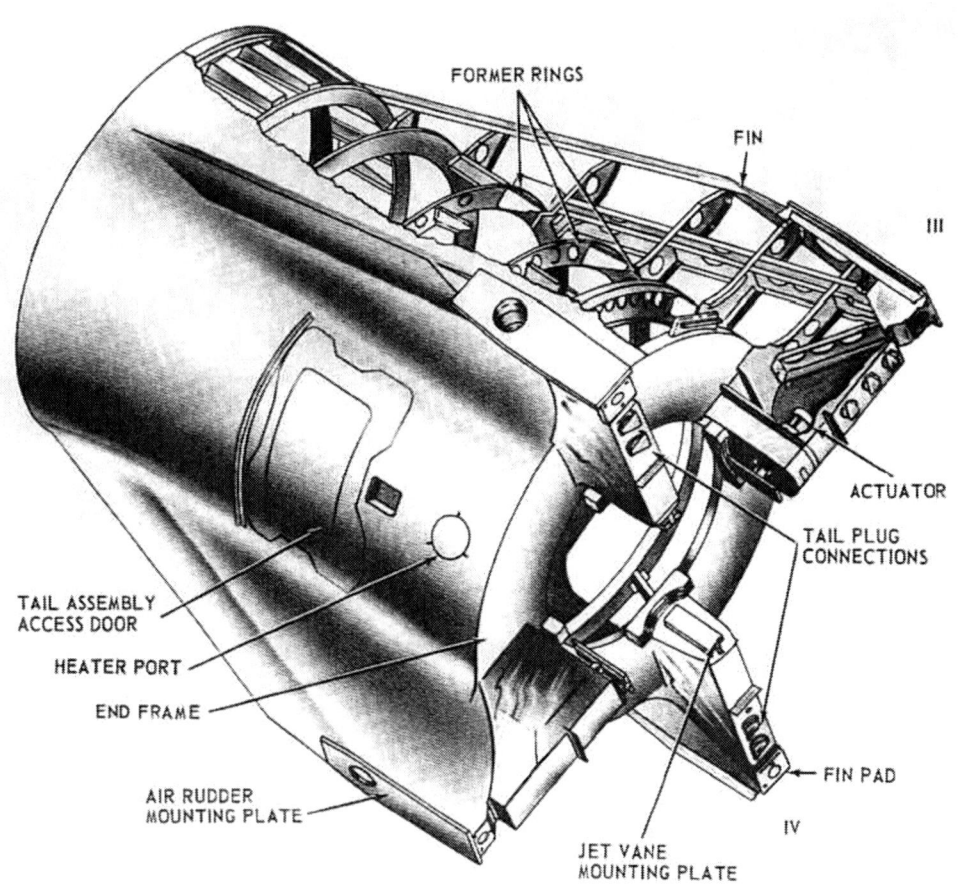

Figure II-7 — Tail Unit

Eight electrical connectors located in the stabilizing fins are used to connect the general electrical network on board the missile with the ground network. After the missile leaves the launcher at liftoff and the connections are broken, spring-loaded caps cover the ends of the connectors to protect them from the engine's exhaust gases.

The emergency cutoff connection is located in fin III.

All fin pads have locating holes for use in mounting the missile on the launcher.

The multiple pneumatic coupling connector is located in a balcony housing on the outside of the tail unit between fins II and III.

The LOX replenishing coupling is located in a similar balcony housing between fins I and IV.

If the missile is to be stored for any length of time, it is necessary to circulate air within the tail unit to prevent the formation of condensation. For this purpose, a hole is provided in the LOX replenishing balcony through which a duct may be inserted for air circulation.

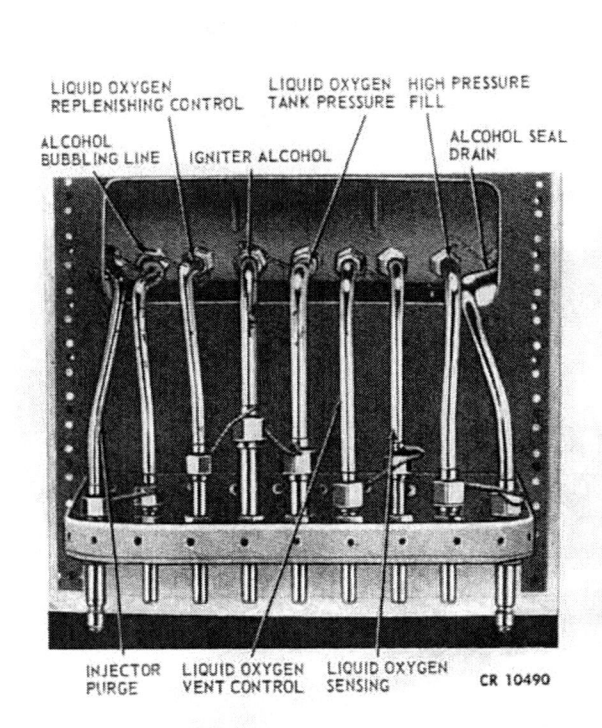

Figure II-8 – Multiple Pneumatic Coupling Balcony

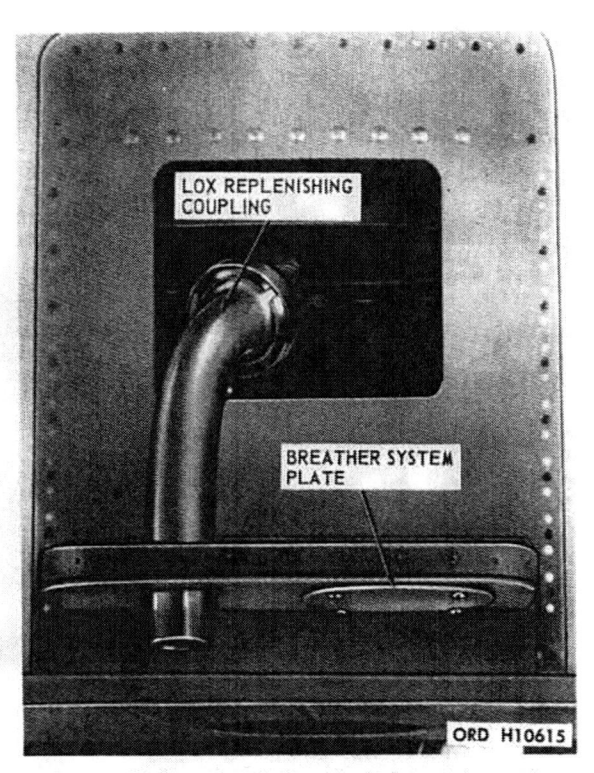

Figure II-9 – LOX Replenishing Coupling Balcony

ENGINE MOUNT

The engine mount is a welded frame of steel tubing that is bolted to the center unit. The mount transmits thrust produced by the engine to the aft ring frame of the center unit. Shims are used between the engine mount and the engine to align the engine. In addition to supporting the rocket engine, the engine mount is used to support the turbopump, the hydrogen peroxide tank, and other components that are related to the propulsion unit.

The engine is of the liquid-propellant, fixed-thrust type consisting of one thrust chamber producing 78,000 pounds nominal thrust. The engine is rated to deliver this thrust for a main-stage duration of 117 seconds. Fuel for the engine is a mixture of alcohol and water with liquid oxygen as the oxidizer. Cooling is accomplished regeneratively by using the alcohol fuel as a coolant. Fuel and oxidizer are delivered to the thrust chamber by a two-stage turbopump which drives the alc and LOX pumps. The turbopump is powered by steam generated through the reaction of hydrogen peroxide and potassium permanganate in the peroxide steam generator. The pumps deliver the propellants at the flow rates and pressures needed to maintain operation.

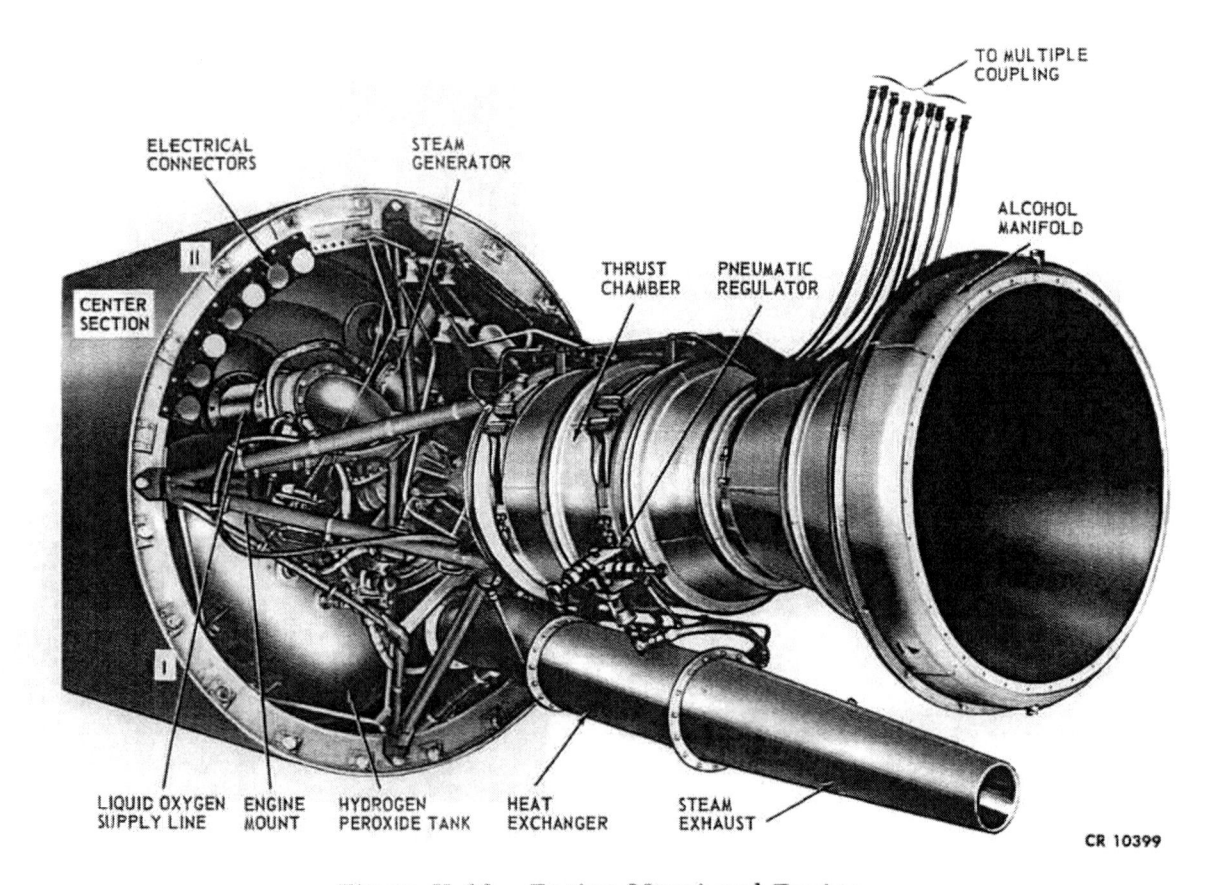

Figure II-10 — Engine Mount and Engine

CHAPTER III
GROUND SUPPORT EQUIPMENT

Figure III-1 — Mobile Launch Site

CHAPTER III
GROUND SUPPORT EQUIPMENT

GENERAL

Ground Support Equipment (GSE) includes all equipment used to transport, handle, test, service, and launch the missile.

One of the major advantages of the REDSTONE Missile System from a tactical and operational point of view is that this system is highly mobile. Mobility was a prime factor in the design and selection of equipment which also had to be rugged, easy to operate, and self-sufficient. The missile and all associated equipment can be transported by air, land, or sea.

Figure III-2 — Loading for Shipment by Air

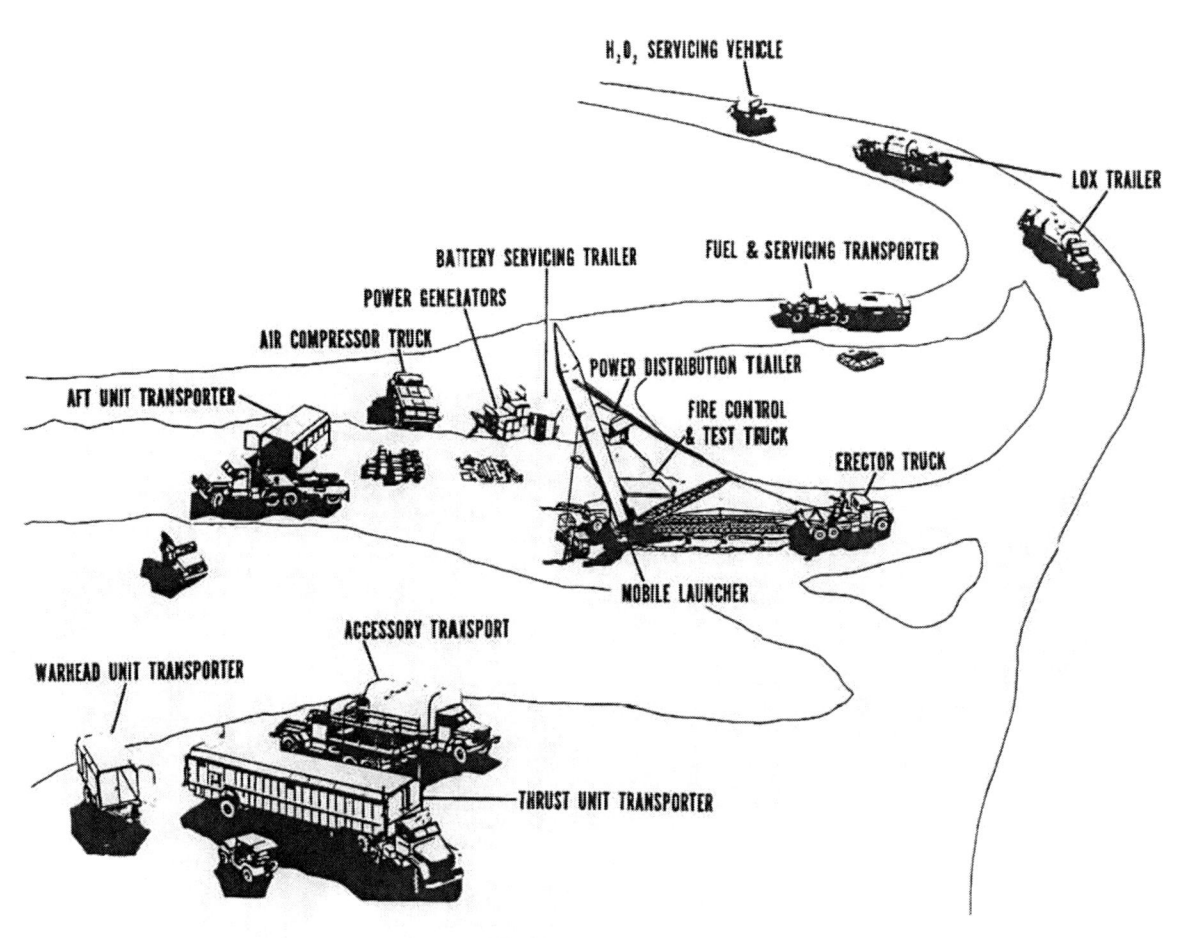

Figure III-3 — Field Deployment

WARHEAD UNIT SEMITRAILER VAN (XM481)

The warhead unit trailer is a two-wheel, single-axle, van-type semitrailer designed to provide transportation and storage for the warhead unit and small on-missile explosive accessories. The weight of the trailer when loaded is 17,660 pounds. It consists of base and cover assemblies and is towed by a 2 1/2-ton, 6 by 6 truck.

Figure III-4 — Warhead Unit Semitrailer (Left Side)

The base assembly is of standard frame construction and contains a rubber-lined saddle, a hold-down clamp, and bolt-receiver supports which secure the warhead unit while in transit. Four jacks equipped with casters are used to lift and remove the cover assembly from the trailer base. Three of the trailers are issued to each Ordnance Support Company and one to each **Artillery Firing Battery**.

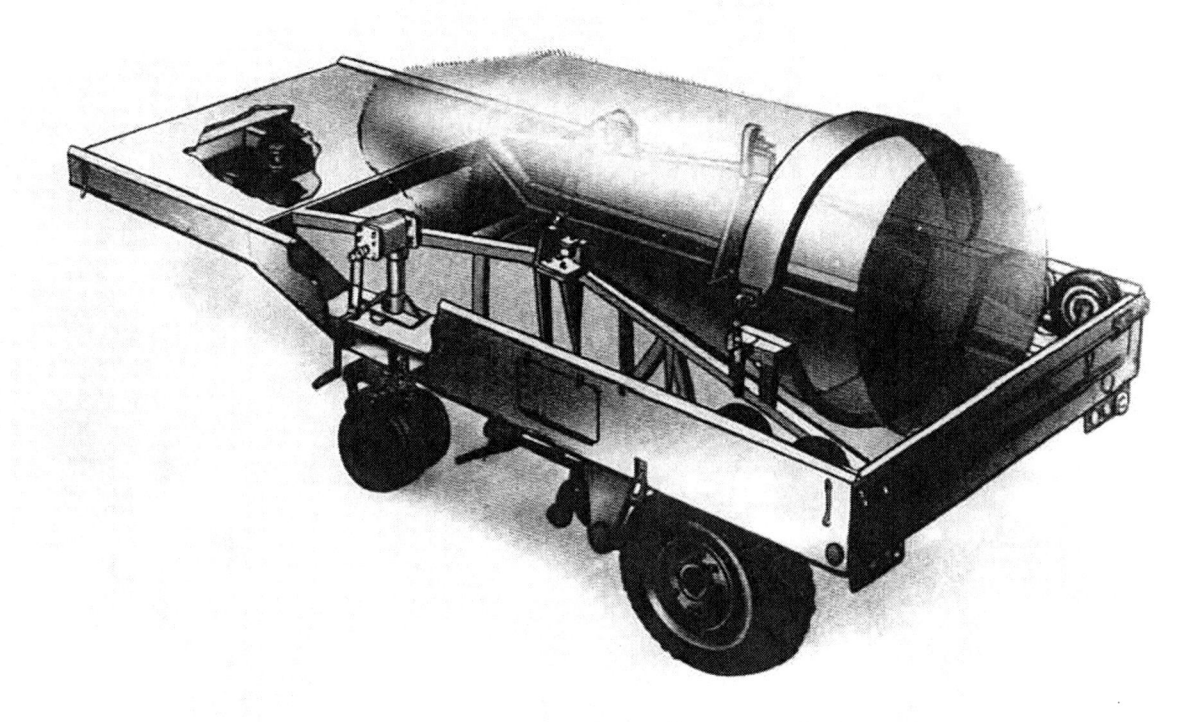

Figure III-5 — Base Assembly — Warhead Unit Trailer

AFT UNIT TRAILER (XM480)

The aft unit trailer is a two-wheel, single-axle, van-type trailer designed to provide transportation and storage for the aft unit and the heater-cooler assembly. This trailer weighs 6,360 pounds when loaded and consists of base and cover assemblies. It is towed by a 2 1/2-ton, 6 by 6 truck.

Figure III-6 — Aft Unit Trailer

The base assembly consists of a standard frame construction that contains a rubber-lined cradle and a hold-down band to secure the aft unit while in transit. The heater-cooler container is mounted on the trailer frame in front of the trailer bed. Four jacks equipped with casters are used to lift and remove the cover assembly from the trailer base.

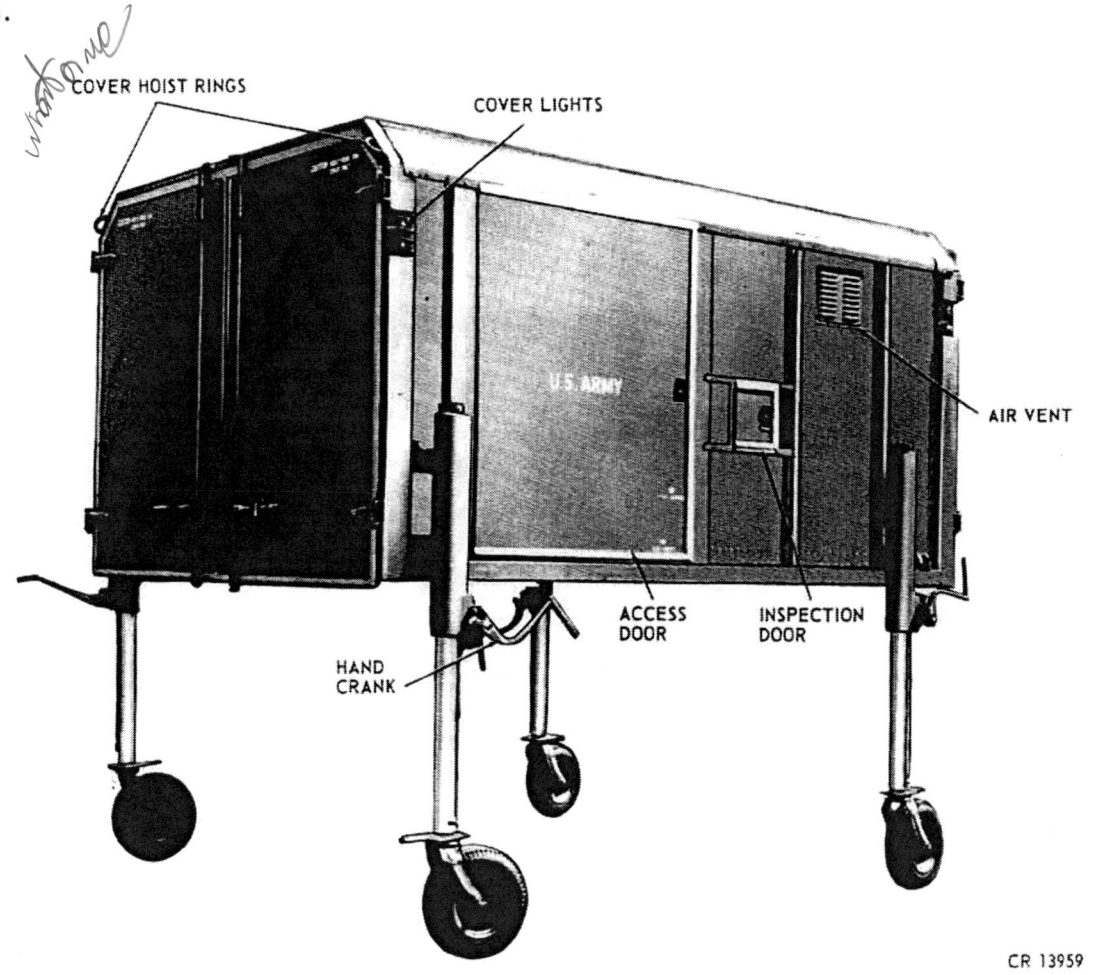

Figure III-7 — Cover — Aft Unit Trailer

III-7

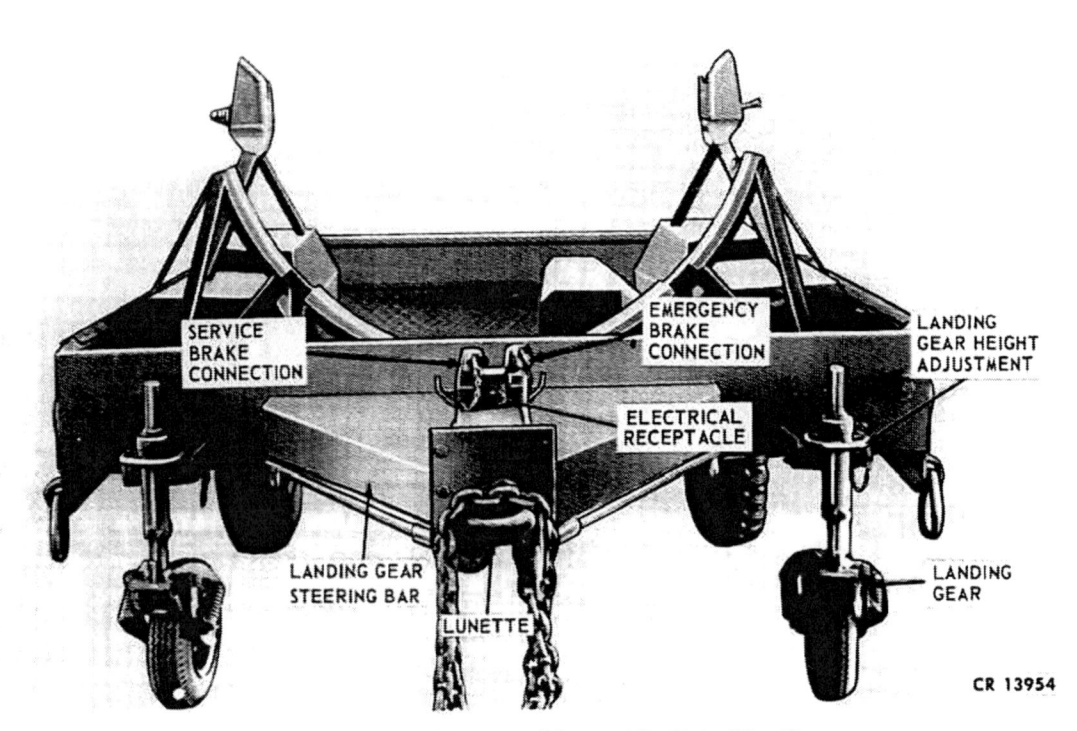

Figure III-8 – Base Assembly – Aft Unit Trailer

III-8

THRUST UNIT TRAILER (XM482)

The thrust unit trailer is a two-wheel, single-axle, van-type semitrailer designed to provide transportation and storage for the missile thrust unit and those missile components which are installed at the firing site. The weight of the trailer when loaded is 18,927 pounds. It consists of base and cover assemblies and is also towed by a 2 1/2-ton, 6 by 6 truck.

The base assembly is of standard frame construction and contains a rubber-lined saddle, a hold-down clamp, and bolt-receiver supports which secure the thrust unit while in transit. Four jacks equipped with casters are used to lift and remove the cover assembly from the trailer base. Three of the trailers are issued to each Ordnance Support Company and one to each **Artillery Firing Battery**.

Figure III-9 — Thrust Unit Trailer

Figure III-10 — Accessories Transportation Truck (Load A)

Figure III-11 — Accessories Transportation Truck (Load B)

ACCESSORIES TRANSPORTATION TRUCK

The accessories transportation truck is a 2 1/2-ton, 6 by 6, M35 vehicle used to transport loose equipment and the accessories necessary for missile assembly, checkout, and service at the launching site. This equipment consists of unpackaged accessories, packaged accessories, cable containers and cable reels. The truck weighs 18,540 pounds when loaded.

Two of these trucks are issued to each Ordnance Support Company and two to each Artillery Firing Battery.

ERECTOR-SERVICER (XM478)

The guided missile erector-servicer is a 2 1/2-ton, 6 by 6 truck modified to store and transport the A-frame, the H-frame, and the associated erecting and servicing equipment. This vehicle tows the platform launcher. It also provides power (through a 10-ton winch and a 1-ton electric hoist) for assembling and erecting the missile, for erecting the servicing platform, and for operating the personnel elevator.

One of these vehicles is issued to each Artillery Firing Battery.

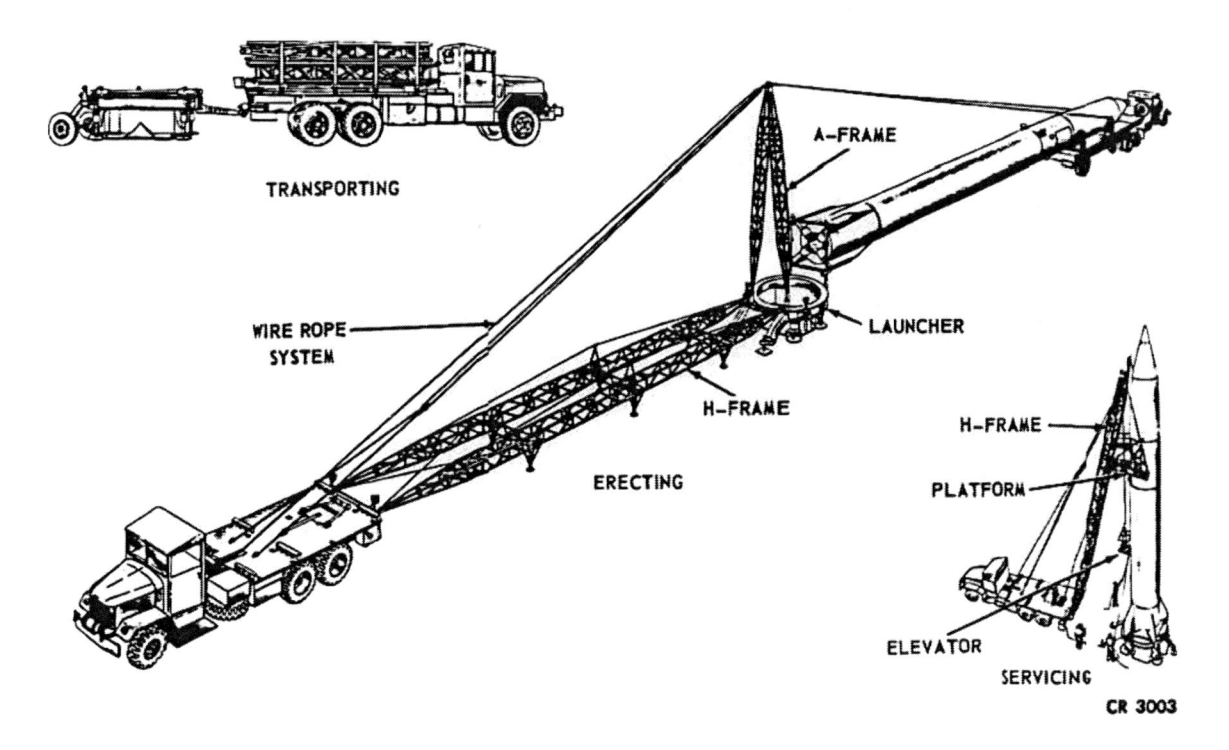

Figure III-12 — Erector Servicer

III-11

MOBILE PLATFORM LAUNCHER (XM74)

The platform launcher is a turntable affixed with two wheels. It is towed by the erector-servicer truck. This launcher provides a base for attaching electrical, pneumatic, and igniter-alcohol connections to the missile, and consists of a base, a deflection plate, a rotating frame assembly, outrigger support arms, and an axle.

Four two-speed jacks are used to support and level the platform launcher. Spirit levels located on the platform launcher at stations II and III are used to level the unit. The rotating frame assembly serves as a hinge during erection and as a support base for launching accessories.

One launcher is issued to each Artillery Firing Battery.

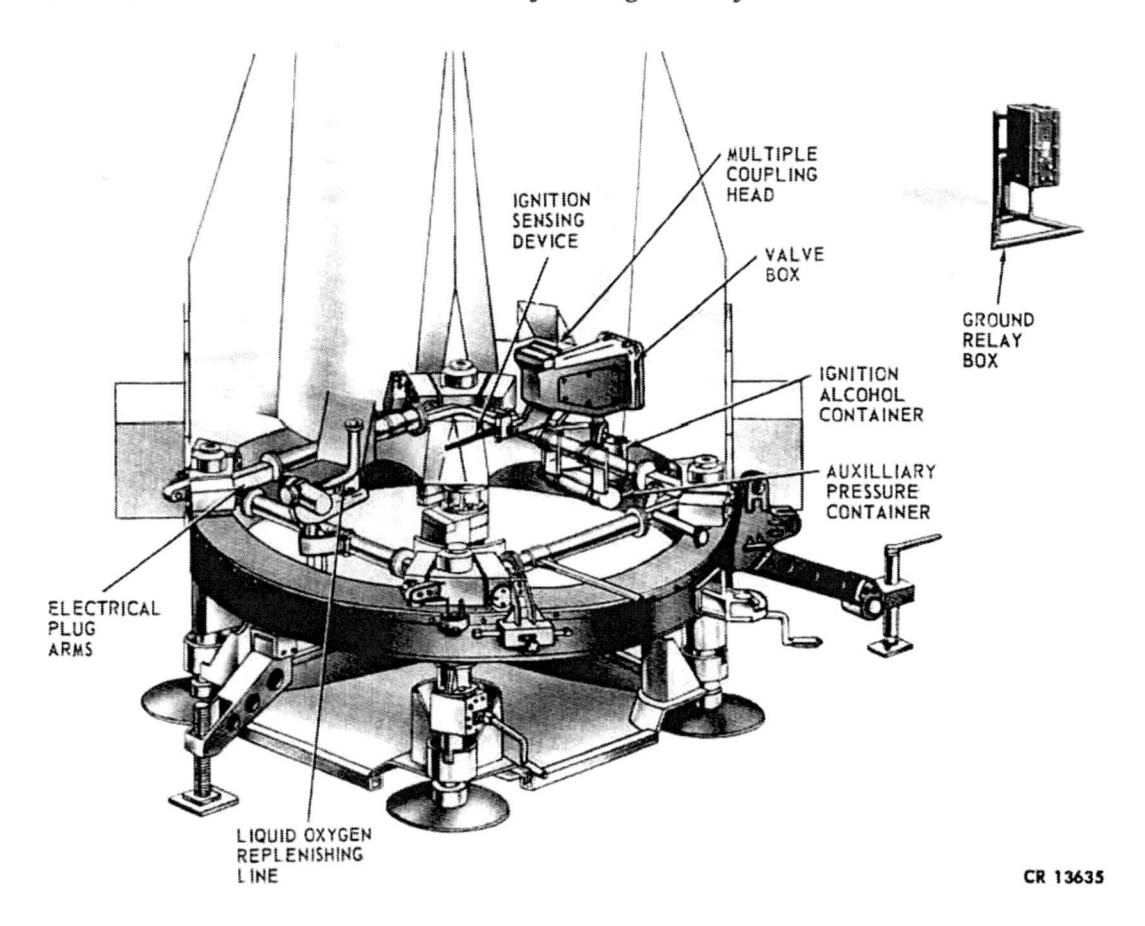

Figure III-13 — Mobile Launcher and Accessories

GUIDED MISSILE PROGRAMER TEST STATION (AN/MSM-38)

The guided missile programer test station is a 2 1/2-ton, 6 by 6 shop van modified to accommodate the personnel and equipment necessary to checkout the REDSTONE Missile. This truck weighs 16,731 pounds. Its primary purpose is to provide mobile facilities for testing certain on-missile components and for preparing the missile for flight by establishing predetermined trajectory data within the guidance system.

Figure III-14 — Guided Missile Programer Test Station (Exterior)

This van is divided into two sections: an operations room and a vestibule. These compartments are separated by a metal partition with a sliding door. At the rear of the van, there is a hinged platform which is used during loading and unloading. A detachable ladder is fixed to the platform for convenient access.

One vehicle is issued to the Ordnance Support Company and one to each Artillery Firing Battery.

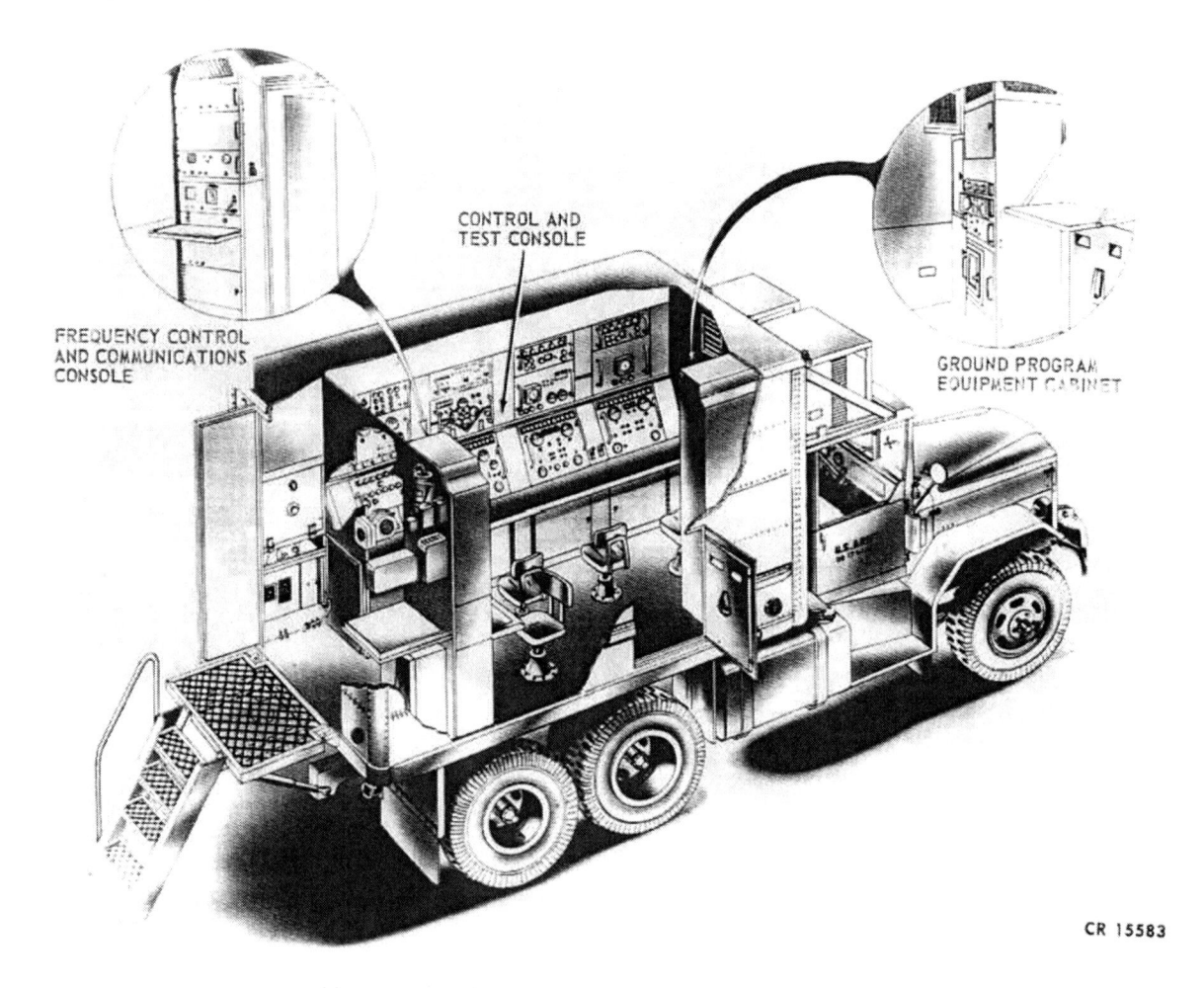

Figure III-15 — Guided Missile Programer Test Station (Interior)

GENERATOR SET

The generator set consists of a portable, diesel-driven, 60-kw generator and accessories mounted on a 3/4-ton, two-wheel trailer. The complete unit weighs 8,050 pounds when loaded. The generator provides power (120-volt, single-phase and 208-volt, 3-phase, 60-cps power) to the power distribution station.

One generator set is issued to the Ordnance Support Company, two to the Engineer Support Company, and two to each Artillery Firing Battery.

Figure III-16 — Generator Set

Figure III-17 — Generator Console Panel

Figure III-18 – Generator Motor Compartment

POWER DISTRIBUTION STATION (AN/MSQ-32)

The power distribution station is a 3/4-ton, two-wheel trailer that weighs 2,400 pounds when loaded. This trailer receives electrical power from the generator trailer, converts this power to usable voltage and frequencies, and distributes the power to the other ground equipment and to the REDSTONE Missile. The body of the power distribution station is so constructed that the upper body shell, sides, ends, and top can be removed as a unit.

Inverters, energizers, a power control console, and a junction box are permanently mounted in the power distribution station.

One station is issued to the Ordnance Support Company and one to each Artillery Firing Battery.

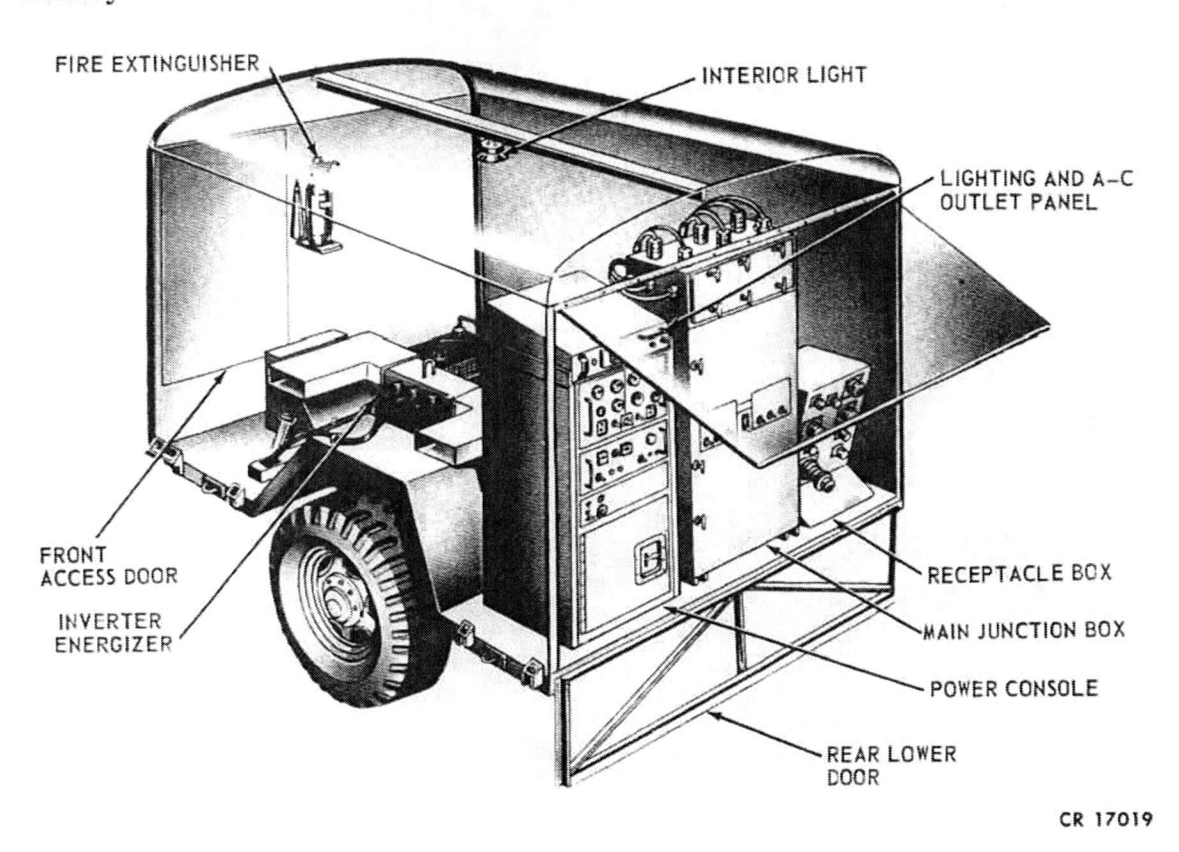

Figure III-19 — Power Distribution Trailer (Left Front View)

III-17

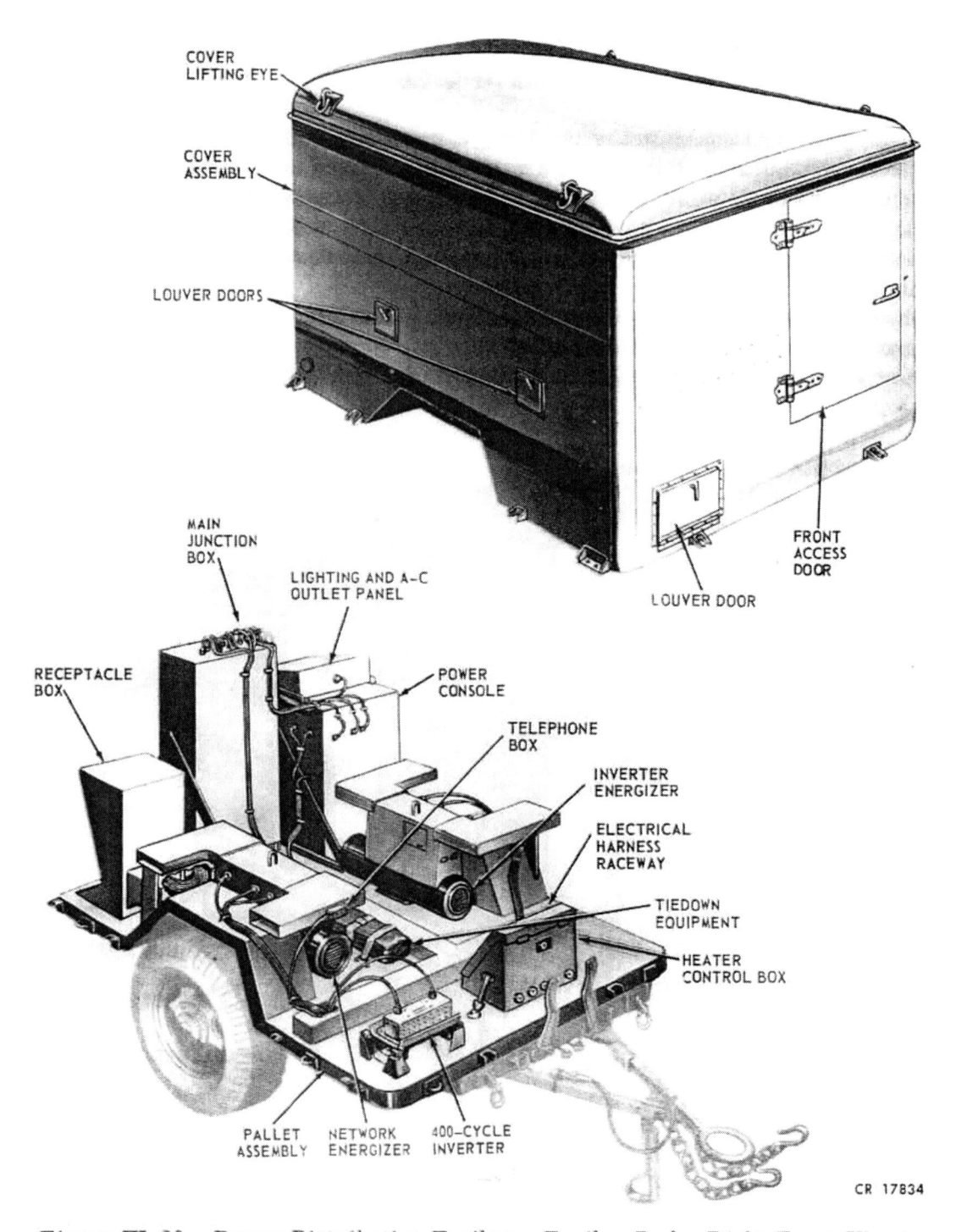

Figure III-20 — Power Distribution Trailer — Trailer Body (Right Front View)

III-18

BATTERY SERVICING SHOP (XM479)

The battery servicing shop is a 3/4-ton, two-wheel trailer designed to store and transport the missile batteries, the battery service and test equipment, and other accessories to the launching site. The battery servicing shop is enclosed by a sheet-metal housing assembly secured to a chassis. The housing assembly contains the equipment used to activate and service the missile batteries prior to launching. The entire unit weighs 2,050 pounds when loaded.

Figure III-21 — Battery Servicing Trailer

DELETED

AIR COMPRESSOR TRUCK

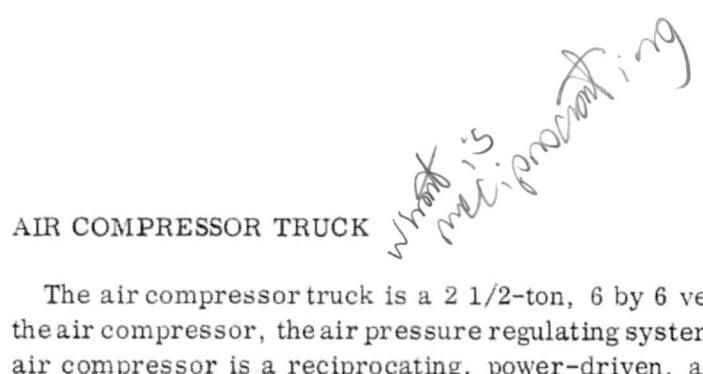

The air compressor truck is a 2 1/2-ton, 6 by 6 vehicle that contains and transports the air compressor, the air pressure regulating system, and accessory equipment. The air compressor is a reciprocating, power-driven, air-cooled unit designed to provide air pressure to the air servicer for testing and pressurizing the REDSTONE Missile. The vehicle weighs 18,540 pounds.

One compressor is issued to the Ordnance Support Company and one to each Artillery Firing Battery.

Figure III-22 — Air Compressor Truck

AIR SERVICER (XM483)

The air servicer is a 3/4-ton, two-wheel trailer designed to function as an air reservoir to assure a continuing air supply at periods of peak demand. The trailer consists of a metal container assembly secured to a chassis. The container assembly houses the air storage battery, air lines, hoses, and miscellaneous components. The complete unit weighs 7,021 pounds.

Figure III-23 – Air Servicer, Right Front 3/4 View

The air storage battery is a self-contained unit secured to a subchassis. Regulators, valves, connecting pneumatic lines, and air bottles are contained in this unit. The pneumatic system includes two panels with mounted controls to vent and monitor the pressurized air.

Issued to Ordnance Support Company and Artillery Firing Battery.

Figure III-23 — Air Servicer, Left Rear 3/4 View

ALCOHOL TANK SEMITRAILER (XM388)

The alcohol tank semitrailer is a two-wheel, 3,000-gallon, tank-type vehicle designed to transport fuel from the storage area to the launching site, and to fill, recirculate, or drain fuel as required at the launching site. The alcohol semitrailer consists of an elliptical, single-compartment tank, an undercarriage, a pumping compartment and a heater system. The fuel-transfer equipment is located at the rear of the tank and provides for metering, filtering, and transferring the fuel to the missile. A 5-ton, 6 by 6, M-52 tractor truck serves as the prime mover for the alcohol semitrailer. The semitrailer weighs 24,300 pounds when loaded.

Three are issued to the Ordnance Support Company and one each to the Artillery Firing Battery.

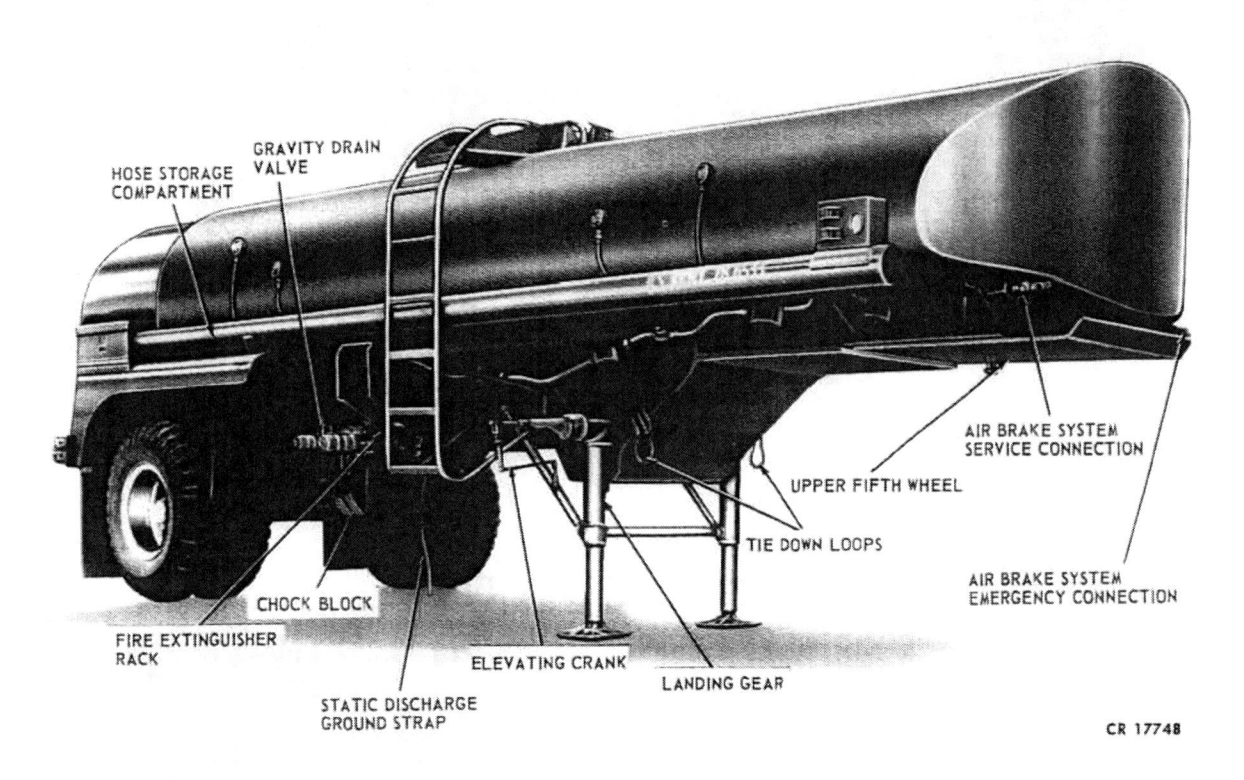

Figure III-24 — Alcohol Tank Semitrailer

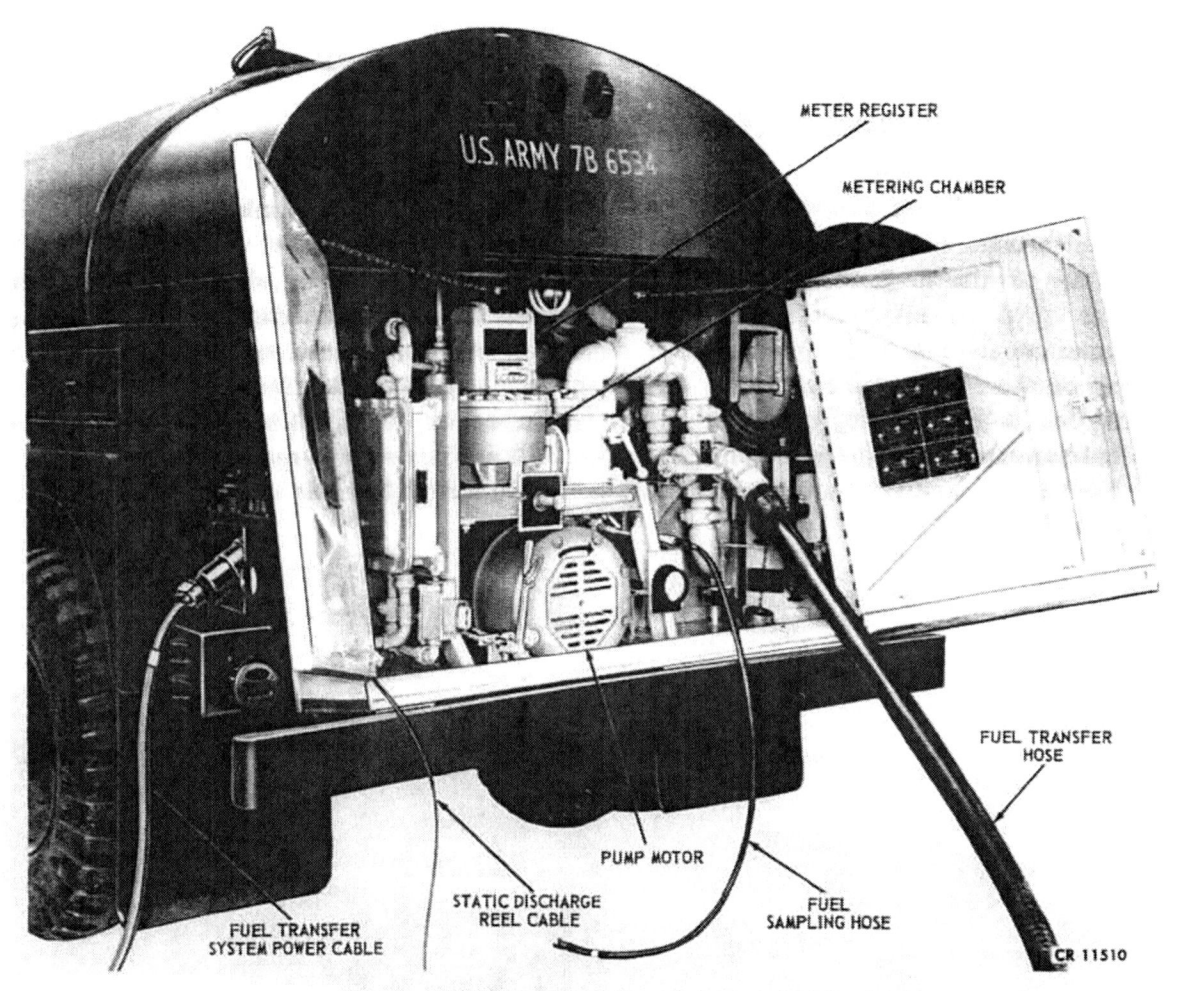

Figure III-25 — Pumping Compartment

AIR SUPPLY SEMITRAILER

The air supply semitrailer is operated in conjunction with the air separation semitrailer to produce LOX and liquid nitrogen for use in the REDSTONE Missile. The air supply semitrailer contains the compressed air supply assembly powerplant, powered by four diesel engines, that drives four 4-stage air compressors. This semitrailer is towed by a standard M-52 tractor truck.

Three semitrailers are issued to the Engineer Support Company.

Figure III-26 — Air Supply Semitrailer

III-24

AIR SEPARATION SEMITRAILER

The air separation semitrailer is operated in conjunction with the air supply semitrailer to produce LOX and liquid nitrogen for use in the REDSTONE Missile. The air separation semitrailer contains the oxygen-nitrogen separation assembly consisting of a heat exchanger, air dryers, refrigeration system, a distillation column and an electrical generator. This semitrailer is towed by a standard M-52 tractor truck.

Three semitrailers are issued to the Engineer Support Company.

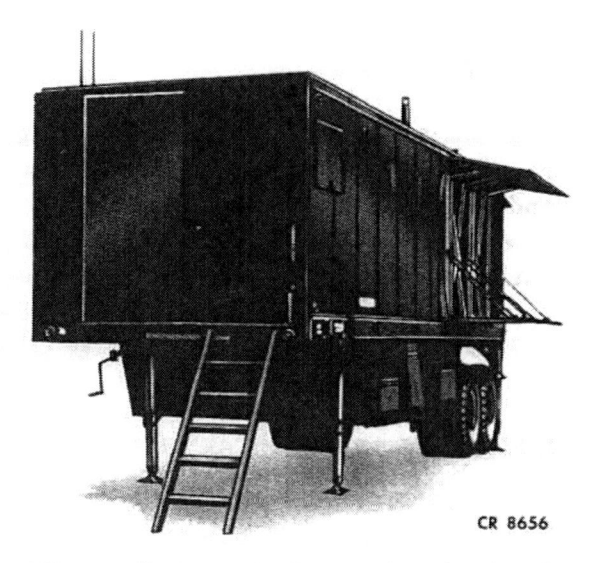

Figure III-27 — Air Separation Semitrailer

LOX TANK SEMITRAILER

The LOX semitrailer is a two-wheel, tank-type vehicle with a LOX capacity of 9 tons. It is designed to transport LOX from the storage area to the launching site, and to transfer LOX to the REDSTONE Missile. The LOX transfer equipment is mounted in a closed compartment at the rear of the tank. Initial filling of the missile tanks at the launching site and final topping to replace evaporation losses are accomplished from the LOX semitrailers. A 5-ton, 6 by 6, M-52 tractor truck serves as the prime mover for the LOX semitrailer. This semitrailer weighs 35,640 pounds when loaded.

Twelve LOX semitrailers are issued to the Engineer Support Company and two to each Artillery Firing Battery.

Figure III-28 — LOX Semitrailer

LOX STORAGE CONTAINER

The LOX storage container is a low-pressure skid-mounted tank with a rated storage capacity of 70,000 pounds of LOX. The container is mounted on two steel I-beam skids and rests on a cradle that is welded to the outer shell of the container. The container is equipped with a pump and with separate fill and discharge connections located on each side of the control cabinet at the rear of the container. The empty container can be transported on a standard 25-ton, low-bed semitrailer.

Two of these containers are issued to the Engineer Support Company.

HYDROGEN PEROXIDE SERVICER (XM387)

The hydrogen peroxide servicer is a modified 3/4-ton, 4 by 4 truck designed to heat, store, transport, and transfer the hydrogen peroxide used in the REDSTONE Missile. A tarpaulin roof, supported by bows, covers the cargo compartment which contains a 78-gallon drum of hydrogen peroxide, a monorail, and chain hoist assemblies. The monorail assembly extends beyond the tailgate and serves as a track for the chain hoist for lifting and lowering the hydrogen peroxide drum. This servicer weighs 7,630 pounds when loaded.

One hydrogen peroxide servicer is issued to each Artillery Firing Battery.

BULK MATERIAL REPAIR PARTS TRUCK (XM486)

The bulk material repair parts truck is a modified 2 1/2-ton, 6 by 6 truck designed to store and transport general repair parts and special bulk items to the launch site. This truck is equipped with a detachable, corrugated, sheet-metal housing assembly, bin assemblies, and a vertical storage rack. The truck also contains the necessary facilities for properly storing odd-size repair parts.

One truck is issued to the Ordnance Support Company and four to the Engineer Support Company.

REPAIR PARTS TRUCK (XM488)

The repair parts truck is a modified 2 1/2-ton, 6 by 6 truck designed to store and transport standard and tray-size repair parts to the vicinity of the launching site. This truck is equipped with a detachable, corrugated, sheet-metal housing assembly, a 4-bay rack assembly, and an access ladder. The truck also contains the necessary facilities for properly storing repair parts carried in bin stock.

One of these trucks is issued to the Ordnance Support Company.

Figure III-29 — Hydrogen Peroxide Servicer

REPAIR PARTS TRAILER (XM487)

The repair parts trailer is a modified 1 1/2-ton, two-wheel trailer designed to store and transport standard and tray-size repair parts to the vicinity of the launch site. This trailer is equipped with a detachable, corrugated, sheet-metal housing assembly and two 6-bay rack assemblies. The trailer also contains the necessary facilities for properly storing repair parts carried in bin stock.

Two are issued to the Ordnance Support Company, three to the Engineer Support Company, and one to each Artillery Firing Battery.

Figure III-30 — Repair Parts Trailer

PRESERVATION AND PACKAGING SHOP (XM485)

The preservation and packaging truck is a modified 2 1/2-ton, 6 by 6 truck designed to store and transport packaging and preservation materials. This truck is equipped with a detachable housing assembly, rack assemblies, a work bench, and storage cabinets. The truck also contains the necessary facilities for properly preserving and storing parts of the REDSTONE Missile.

One truck is issued to the Ordnance Support Company.

Figure III-31 — Basic Truck (Bulk Material Repair Parts, Repair Parts, Preservation Packaging)

PNEUMATIC SHOP (XM477)

The pneumatic shop is a 2 1/2-ton, 6 by 6 truck modified to transport and store the equipment required to test and repair pneumatic systems components. The pneumatic shop contains a checkout bench consisting of an electrical and pneumatic compartment, a control well, a holding-fixture test well, and drawers for storing special test fixtures and fittings. The electrical and pneumatic compartment contains components which provide electrical power and air pressure to the control panel during test operations.

One truck is issued to the Ordnance Support Company.

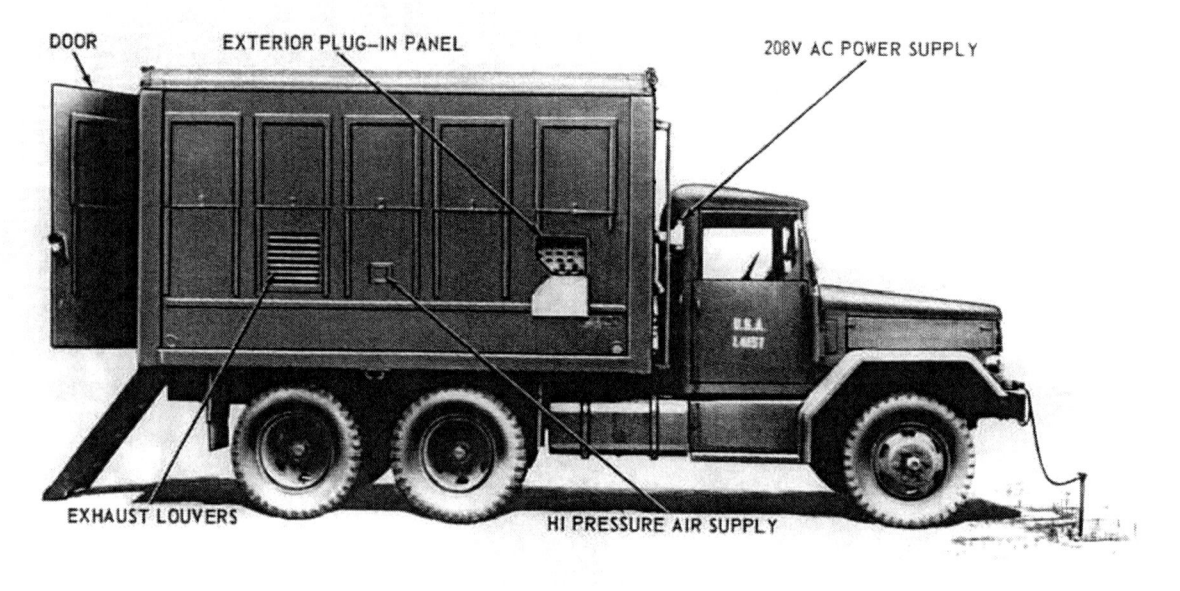

Figure III-32 — Pneumatic Shop Truck

SUPPLY OFFICE (XM484)

The supply office is a 3-ton, two-wheel, semitrailer designed to store and transport the necessary files and clerical equipment used to maintain repair parts records for the REDSTONE Missile. Located along one wall of the supply office is a bookcase, a repair parts record file, a repair basket rack, a graph-o-type embossing machine, and a heater compartment. Along the opposite wall are four Kardex file cabinets, a bookcase, an Addressograph plate storage cabinet, and a classified file.

One supply office is issued to the Ordnance Support Company and to the Engineer Support Company.

Figure III-33 — Supply Office and Prime Mover

STABILIZED PLATFORM TEST STATION (AN/MJM-2)

The stabilized platform test station is a 3-ton, two-wheel van designed to transport and house equipment for accurately testing the ST-80 stabilized platform assembly. The equipment for testing the ST-80 stabilized platform is located in two compartments. The front compartment contains a pedestal test stand, an accuracy test console, the air supply unit, a power panel, and amplifier racks. The rear compartment contains a preset and warmup console, the ST-80 warmup test cart, a power components cabinet, and amplifier racks.

One van is issued to the Ordnance Support Company.

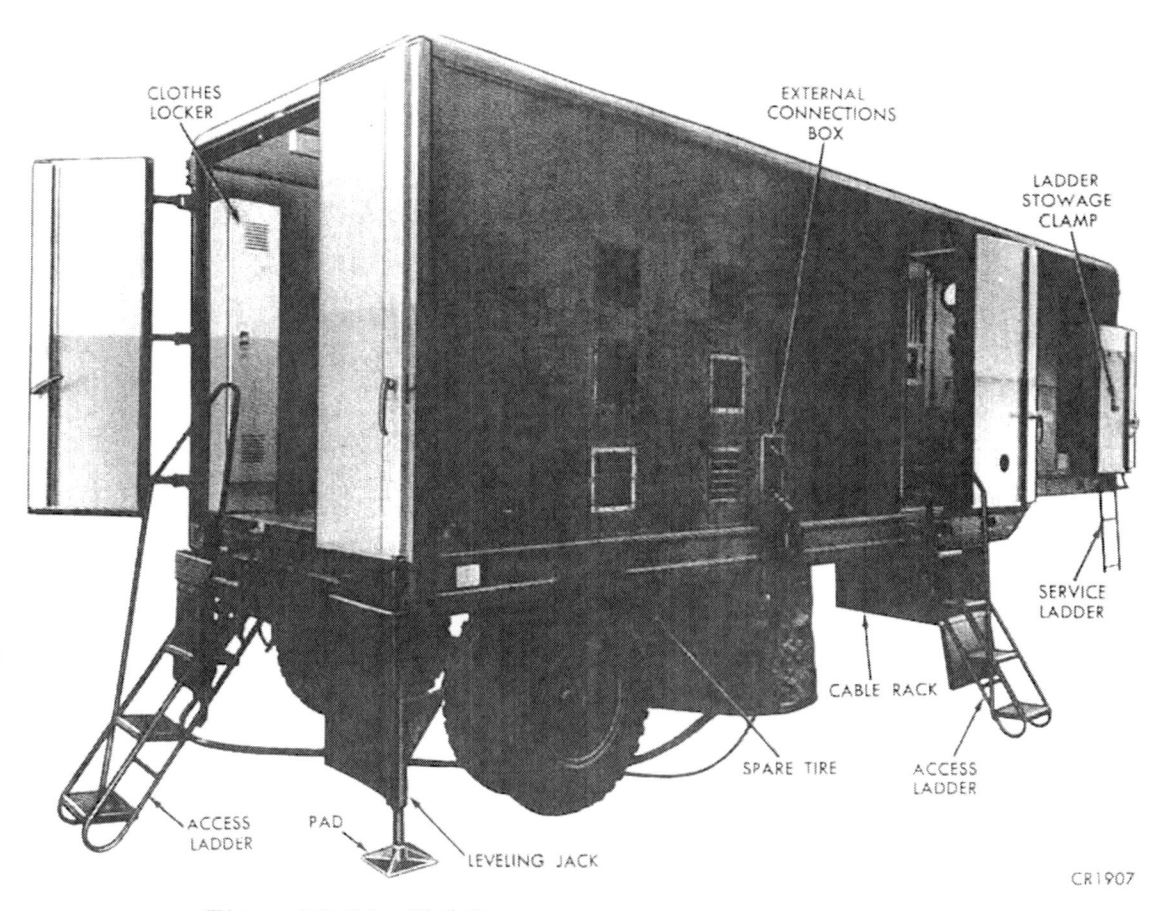

Figure III-34 — Stabilized Platform Test Station (Exterior)

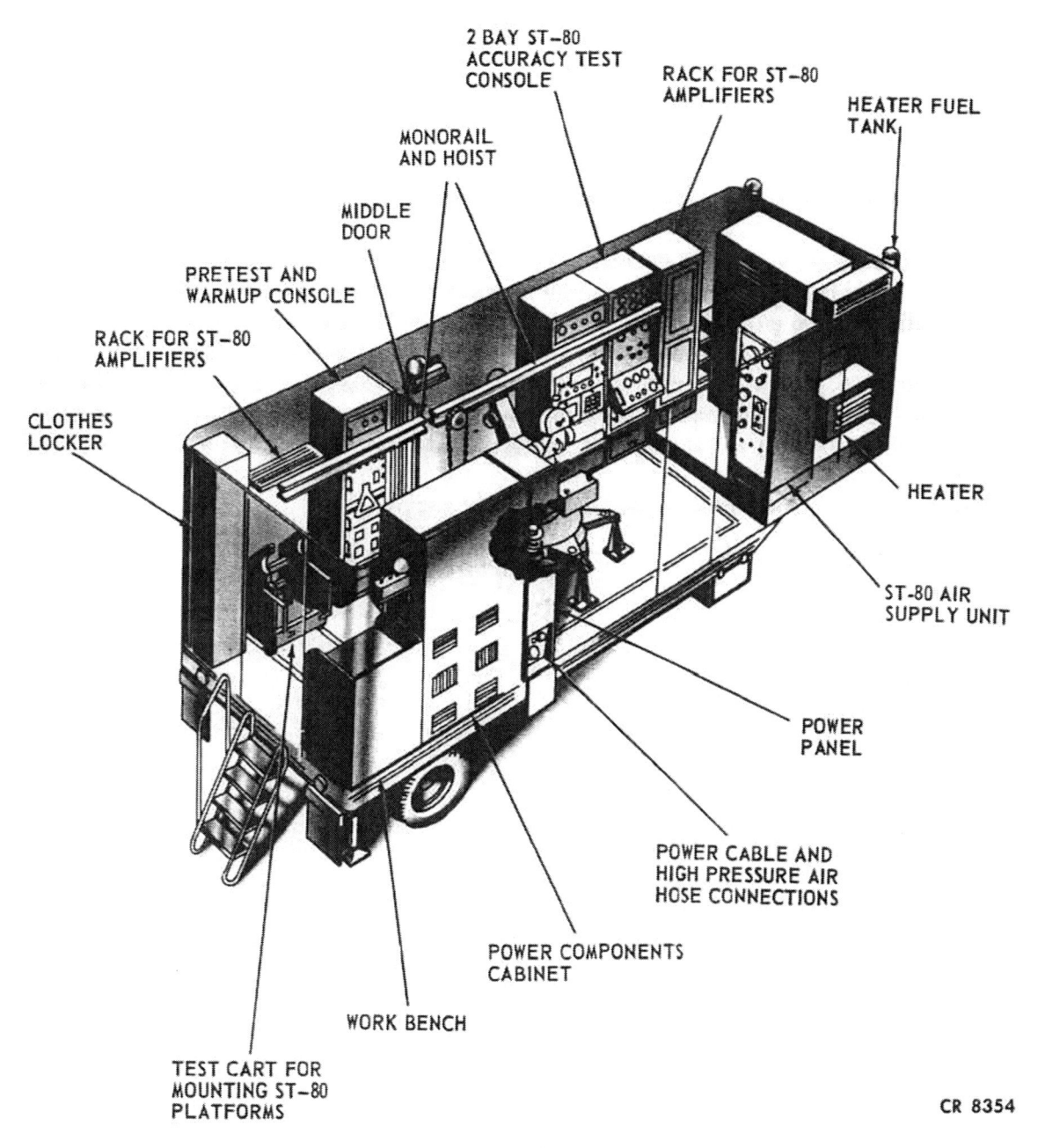

Figure III-35 — Stabilized Platform Test Station (Interior)

GUIDANCE AND CONTROL COMPONENTS TEST TRAILER "A"

The guidance and control components test trailer "A" is a 3-ton, two-wheel van designed to transport and house the ST-80 turn-tilt stand and the equipment required to functionally test the guidance computer, the program device, and the ST-80 stabilized platform. The guidance computer enclosure unit is located at the forward end of the electronic shop and is equipped with a heater and fan, mounts, cables, and cable receptacles. The auxiliary equipment contained in the electronic shop includes a monorail assembly, a trolley and chain hoist, cable reel storage boxes, and equipment access doors.

One trailer is issued to the Ordnance Support Company.

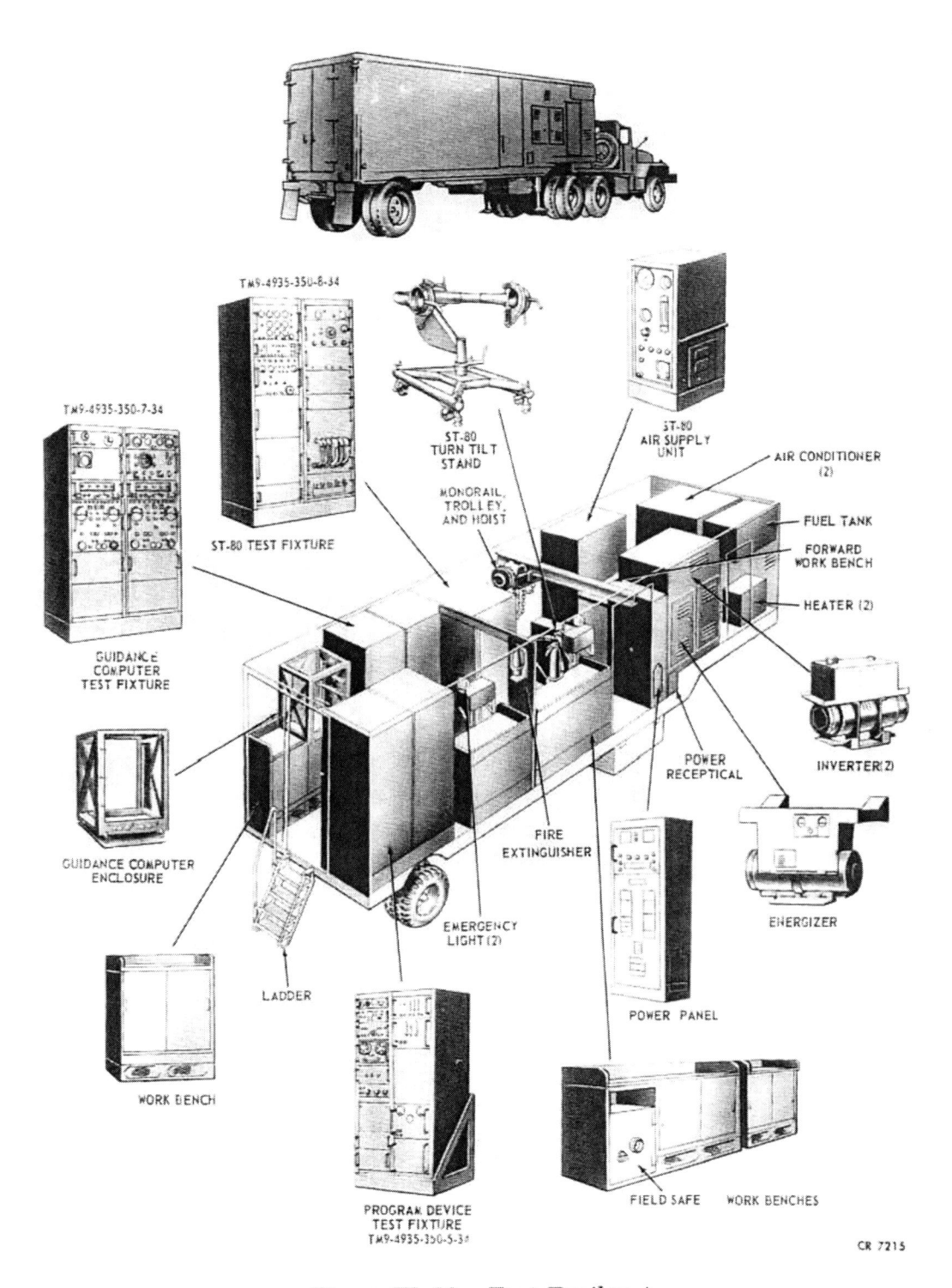

Figure III-36 — Test Trailer A

III-35

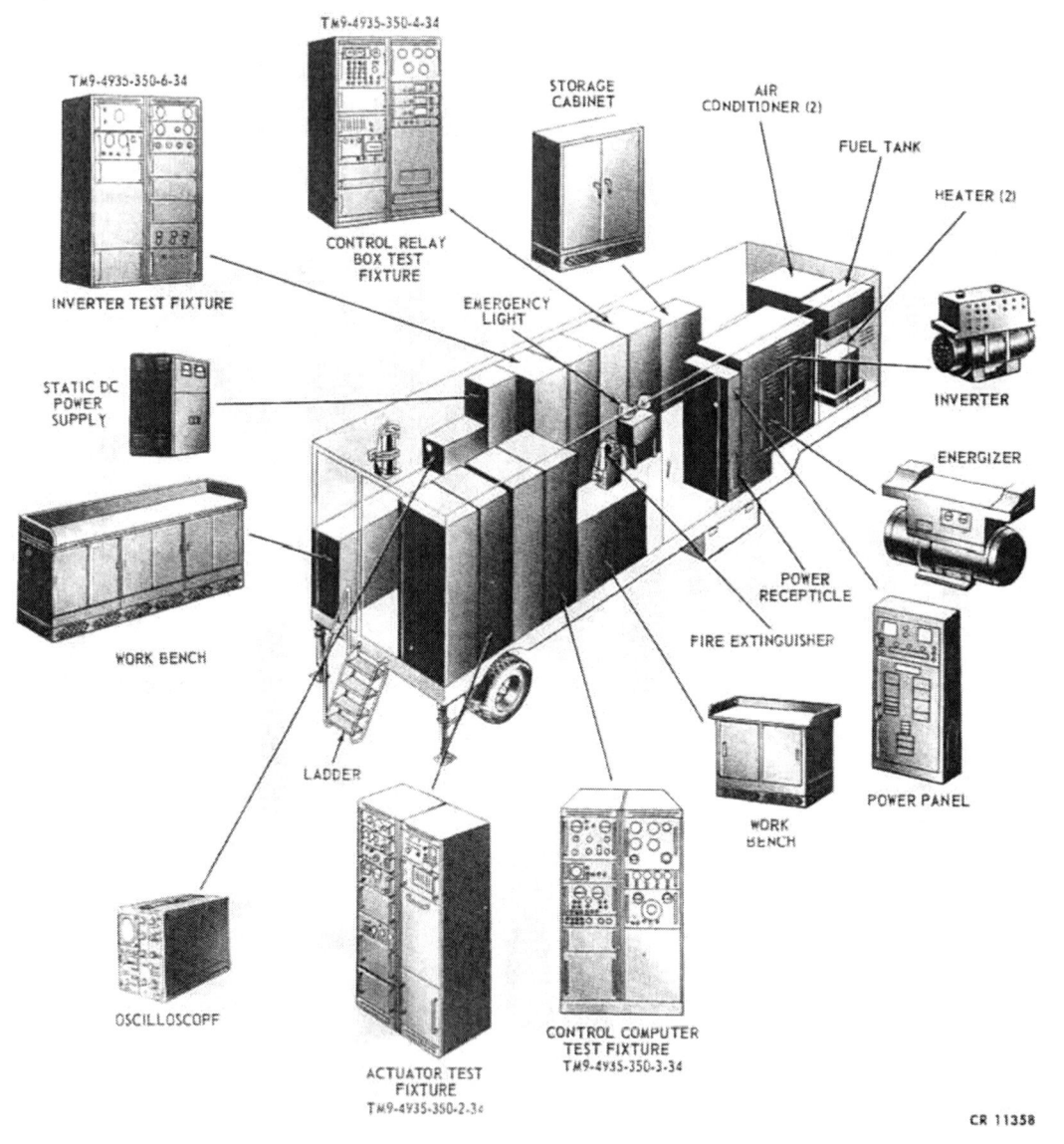

Figure III-37 — Test Trailer B

III-36

GUIDANCE AND CONTROL COMPONENTS TEST TRAILER "B"

The guidance and control components test trailer "B" is a 3-ton, two-wheel van designed to transport and store the equipment required to functionally test the control computer, the control relay box, the actuator, and the inverter. A power panel and receptacle, an energizer, an inverter and a static d-c power supply are contained in the trailer to supply power and perform tests on the related components. The auxiliary equipment contained in the trailer includes air conditioning and heating systems, cable reel storage boxes, and equipment access doors.

One trailer is issued to the Ordnance Support Company.

Figure III-38 — Guidance and Control Test Trailer (A or B) and Prime Mover (M52)

CALIBRATION SET – MISSILE SYSTEM
TEST EQUIPMENT

The calibration vehicle is a 2 1/2-ton, 6 by 6 shop van modified to accommodate the equipment required to properly calibrate all ground support equipment instrumentation. The cargo area of this vehicle incorporates specially constructed racks and benches to house the calibration equipment and to provide maximum protection during transit. The purpose of the calibration equipment is to assure that every meter, gage, and timer in the ground support equipment will function properly and within tolerance.

One vehicle is issued to the Ordnance Support Company.

Figure III-39 – Calibration Set, Missile System Test Equipment

FIRE-FIGHTING EQUIPMENT SET

The fire truck and the auxiliary water tank trailer are provided to fight operational and brush fires. The fire truck is equipped with a powered pump and fire-fighting turret and also the items necessary to combat fires involving fuel and oxidizers used in the REDSTONE Missile. The 2,000-gallon water tank trailer serves as an auxiliary water source for the fire truck.

GUIDED MISSILE TRAINER, ANALYZER VAN

The analyzer van is a 2 1/2-ton, 6 by 6 vehicle used in the REDSTONE training program. The analyzer van contains simulation equipment which furnishes a complete set of responses normally produced by components in the missile, including traceable malfunction responses. The analyzer van also houses an automatic printer and tape recorder system which records both switch operations and personnel conversations during missile checkout.

One van is issued to the Ordnance Support Company.

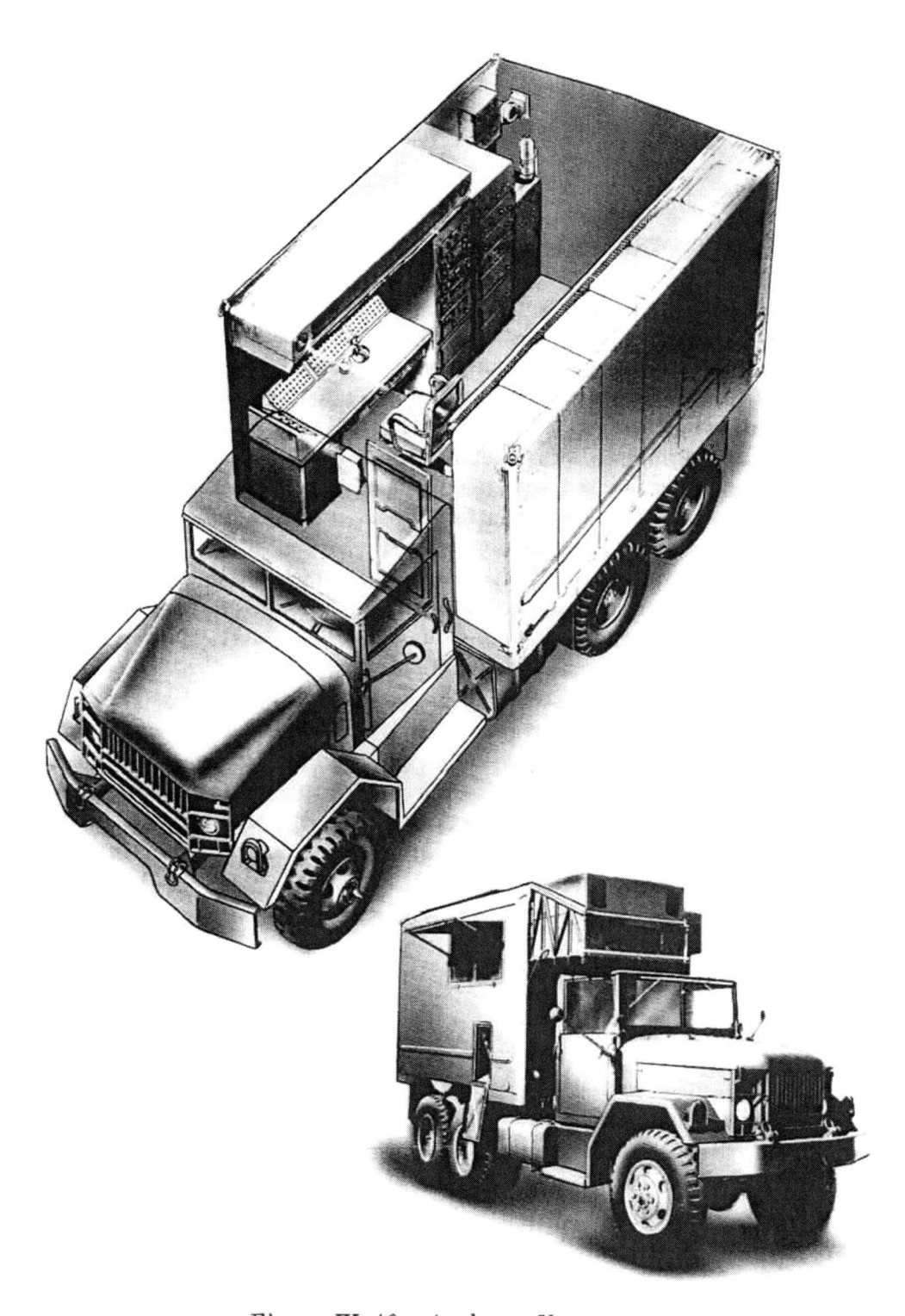

Figure III-40 — Analyzer Van

GUIDED MISSILE TRAINER (AN/MSQ-T2)

The guided missile trainer is a full-scale replica of the REDSTONE Missile. It has been designed to serve as a realistic, long-life trainer which is assembled, checked out, and maintained in the same manner as the tactical REDSTONE Missile. Most of the internal components are duplicated. Inoperative or dummy components are used where operative components are not required for simulation purposes.

Like the tactical missile, the training missile is constructed in four units. All external fueling and pneumatic connections are duplicated. Some of these connections are operative, while others are dummies, but all are capable of receiving the connectors used to service the actual missile.

The warhead unit contains only ballast to give the training missile the proper center of gravity. The aft unit consists of an instrument compartment containing a dummy ST-80 stabilized platform and dummy guidance components. Most of the components contain elements of the traceable-malfunction circuits.

The thrust unit contains a dummy rocket engine, simulated propellant tanks, and pneumatic and electrical systems. LOX is contained in the training missile in a belt tank constructed by placing a second layer of skin 2 1/2 inches inside the outer skin. A 72-gallon tank equipped with overflow lines is utilized for hydrogen peroxide fill training. Training in alcohol fueling is accomplished by using actual fill and vent fittings. In place of an alcohol tank, an external line returns the pumped alcohol to the alcohol semitrailer.

One missile trainer is issued to the Ordnance Support Company.

ST-80 CONTAINER

The ST-80 container is a skid-mounted, drum-shaped container, 49 inches in height and 52 inches in diameter. It is fitted internally with eight coil springs which suspend a handling frame that assures minimum shock and vibration of the ST-80 during shipment. The container has external receptacles to permit pressurization and heating.

The ST-80 is transported to the launch site on an M-105 3/4-ton trailer which is pulled by a prime mover designated by the firing battery commander.

Two containers are issued to the Ordnance Support Company and one to each Artillery Firing Battery.

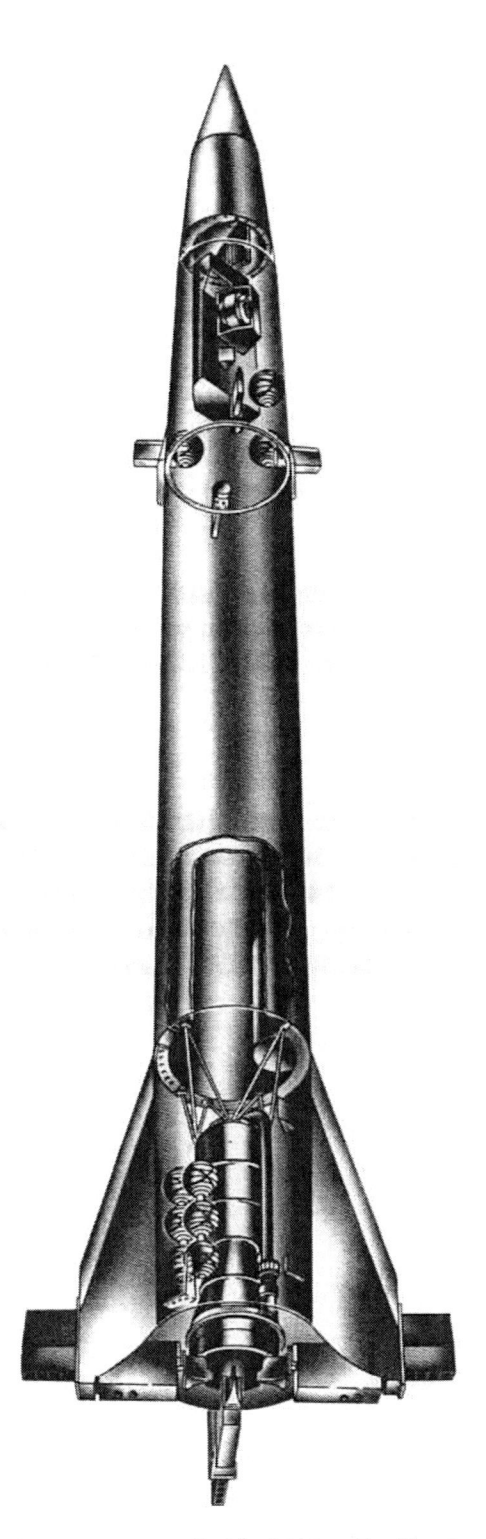

PLEXIGLASS WINDOW- BSP239A

CAP SCREW- MS35298-113

CR 10745

Figure III-41 — Guided Missile Trainer

Figure III-42 — ST-80 Container

III-41

LIQUID NITROGEN CONTAINER

The liquid nitrogen container is a cryogenic tank that has a capacity of 150 gallons. Its function is to supply liquid nitrogen to the heater-cooler system to cool the missile instrument compartment. It is transported to the launch site on a 2 1/2-ton M-35 truck.

MISSILE LAYING KIT

The purpose of the laying kit is to orient the missile to permit the lateral guidance system to be keyed to the theoretical ballistic path from launcher to target. The following equipment is required at the launch site: two Wilde T-2 theodolites with tripod and accessories, one Wilde T-2 precise traverse target, one pocket transit, one 30-meter steel tape, and two surveyor's umbrellas. This equipment is transported on a 1/4-ton, 4 by 4, utility truck.

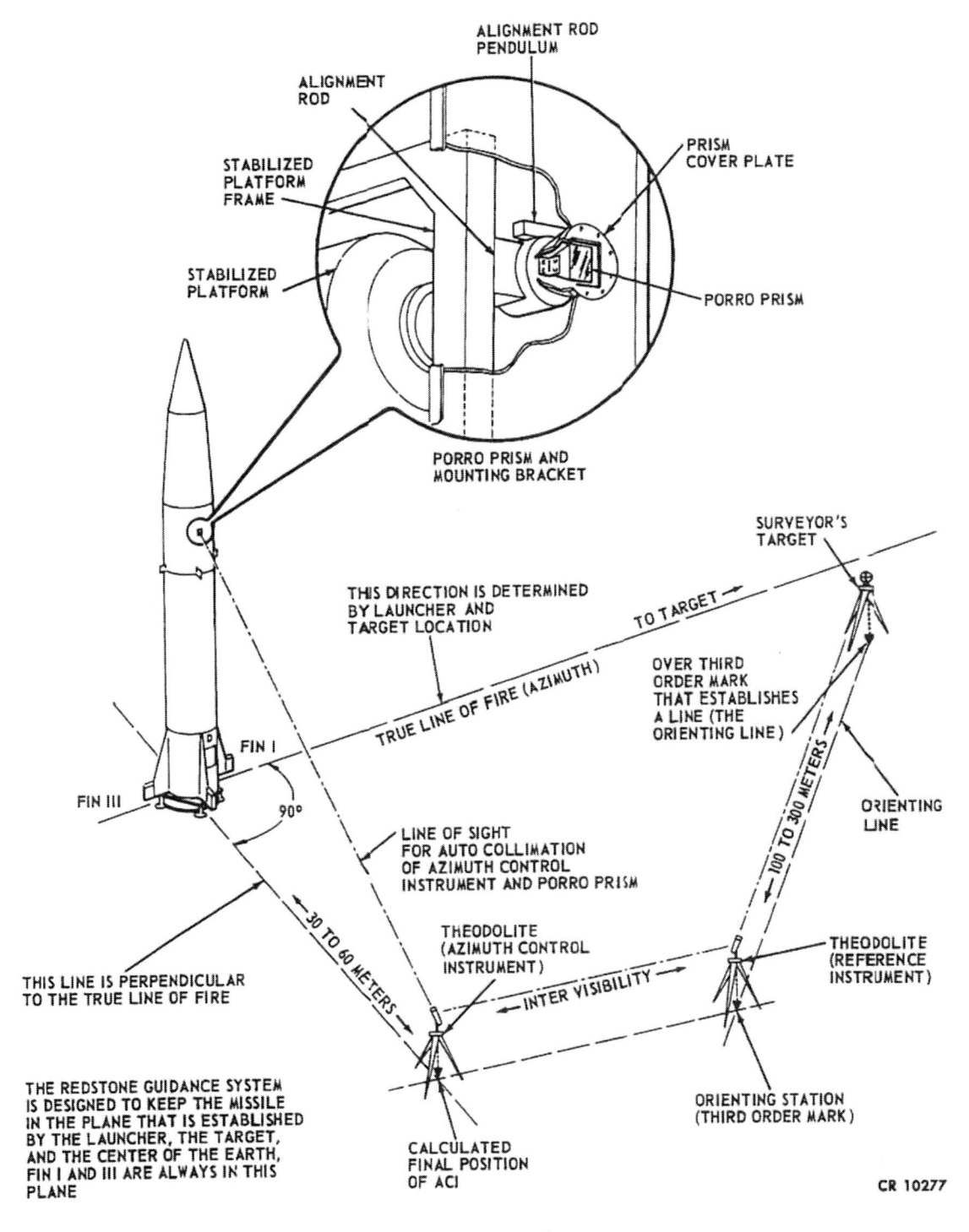

Figure III-43 — Missile Laying

III-43

This page has been left blank intentionally.

CHAPTER IV
GUIDANCE AND CONTROL

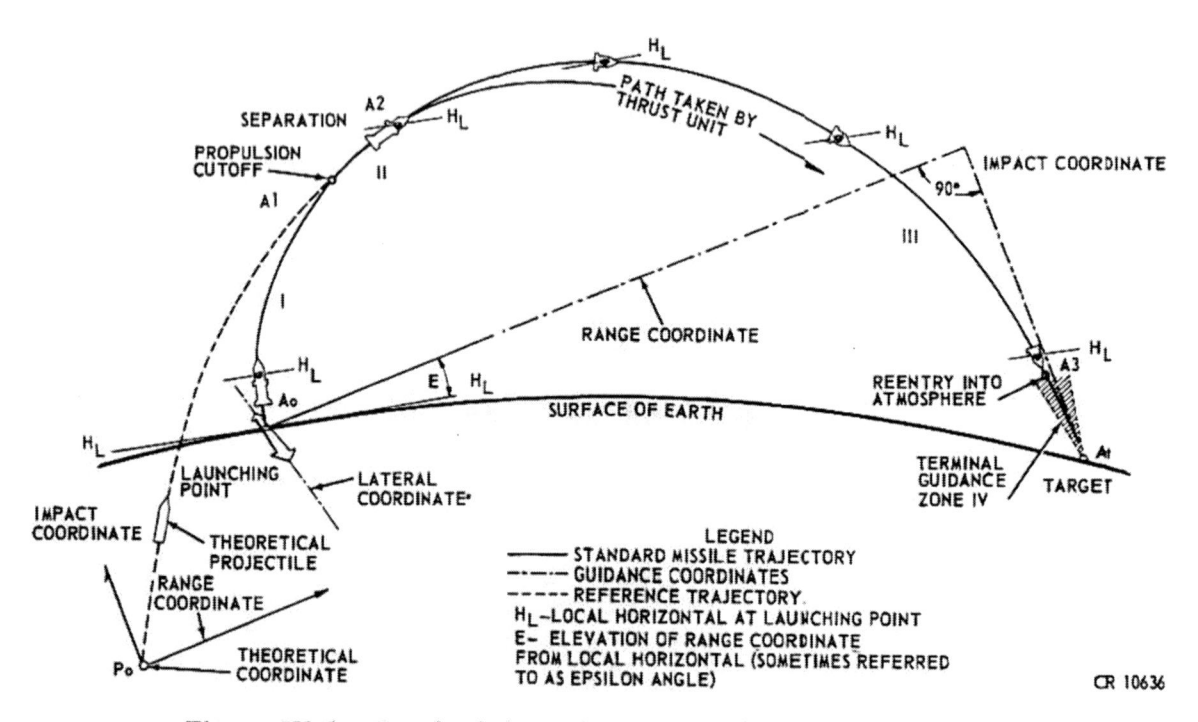

Figure IV-1 — Standard Trajectory and Reference Coordinates

CHAPTER IV
GUIDANCE AND CONTROL

GENERAL

The REDSTONE Missile system employs three trajectories in order to place the warhead on target. The first trajectory to be considered is known as the reference trajectory. This is a pure ballistic trajectory drawn from the target back to a point of launch. The problem then arises that in order to fire the missile from this point and to have it follow a pure ballistic trajectory, two conditions must exist: 1) the missile would have to be fired from an angle other than vertical, and 2) the missile would have to have its cutoff velocity at point of launch. Since neither of these conditions can be met, a practical solution must be found. The solution is to move the launching point closer to the target. The missile may then be fired from a vertical position and programed or tilted into the reference trajectory. If the missile velocity, as the missile enters the reference trajectory, is the same as that of the theoretical missile velocity at that point, the end result will be the same; both missiles would strike the target. This second trajectory is known as the standard trajectory and is the one that is used.

The third trajectory is the actual trajectory: the one the missile follows, including all deviations and corrections. The perfect flight would be one in which the standard and actual trajectories coincided at all points. This perfect flight, however, would never occur because of variations in drag, thrust, winds, and other influencing factors. The function of the guidance and control system is to make the actual and standard trajectories coincide at impact.

The standard trajectory is divided into four phases of flight. Phase I is the portion of flight that extends from launch to cutoff. From launch the missile rises vertically and is then tilted to intercept the reference trajectory. During the first portion of Phase I, the missile is controlled by four carbon vanes located within the jet blast. When the missile has reached a velocity sufficient for it to become aerodynamically stable, four air rudders located at equidistant points around the outside of the tail section take over the control function. The transition of control from carbon vanes to air rudders, however, is gradual because both the vanes and the rudders are driven by the same actuator and both turn simultaneously. At cutoff, the missile will be out of the earth's atmosphere and will have sufficient velocity to carry it to the target. Cutoff will occur between 96 and 107 seconds after launch, depending upon the range. Phase II begins with engine cutoff and ends with the separation of the body unit from the thrust unit; missile separation occurs at 127 seconds after launch.

Phase III begins at separation and ends at a point called "Q" (the point at which the missile re-enters the earth's atmosphere). This portion of flight takes place outside of the earth's atmosphere and is therefore known as spatial flight. During spatial flight, missile control must be accomplished by an action-reaction type device, since any aerodynamic device would be useless because of the lack of lift. These control devices are air jets which are located around the circumference of the skirt section. Phase IV is from "Q" to impact and is generally referred to as the "dive phase".

The REDSTONE, like any other missile, has six degrees of freedom: three angular and three translatory. The missile is free to pitch and yaw about its center of gravity and to roll about its longitudinal axis. It is also free to be displaced to the right and left, up and down, or backwards and forwards. The three angular movements deal with the missile's attitude and are, therefore, control functions; the three translatory movements deal with the missile's physical displacement and, therefore, are guidance functions. This is the distinction between guidance and control. The REDSTONE guidance and control system is capable of measuring any deviation in attitude or any displacement from the standard trajectory. This is accomplished by the use of a stabilized platform known as the ST-80. The ST-80 provides the missile with a space-fixed frame of reference. The platform is stabilized at the launching site and will maintain a reference, throughout flight, space-fixed to the local horizontal at the time of launch. By placing potentiometers between the stable platform and the missile's airframe, a means is provided for detecting and measuring the angular rotation of the missile about this reference. By using three potentiometers, errors in the pitch, roll, and yaw planes are detected and measured. These error signals carry the designation of ϕ (phi) signals and are classed as attitude error signals. With proper amplification and distribution to the various control devices, these signals are used to control the missile attitude during flight. A means is provided whereby the reference point of the pitch control potentiometer or command potentiometer may be changed during flight to cause the missile to pitch over and follow a ballistic trajectory. This is done through the use of a Program Device that feeds a continuous series of pulses to the stabilization system, causing the zero position of the pitch command potentiometer to shift. The missile control system will recognize this shifting zero point as an attitude error signal and will cause the missile to tilt over until the wiper on the potentiometer is aligned to the new zero. This process is known as pitch programming and is the means by which the missile is made to enter a ballistic trajectory and to maintain correct pitch attitude throughout flight.

In order to sense its physical position in relation to the standard trajectory, the REDSTONE Missile carries two gyro accelerometers: one for lateral measurements and one for range measurements. The lateral accelerometer measures accelerations to the right or to the left of the trajectory, and the range accelerometer measures the acceleration along the range coordinate. By first integration of these accelerometer signals, velocity information is obtained; second integration provides displacement information. First integration takes place within the accelerometer unit itself. The accelerometers are mounted on the ST-80 because this is the only component within the missile that does not change its orientation during flight. The sensitive axes of both accelerometers must be aligned along their measuring coordinates. The lateral coordinate is crosswise to the trajectory plane. The range coordinate is formed by drawing a line from the point of launch to intersect at an angle of 90 degrees, a line drawn tangent to the trajectory at point of impact. The angle formed by the range coordinate and the local horizontal is the epsilon (ϵ) angle and varies from 20 to 43 degrees. The range accelerometer is aligned to this angle prior to launch.

As shown in the guidance and control block diagram (Figure IV-2), the REDSTONE guidance and control system is made up of five basic subsystems: Stabilization, guidance, control, program, and power.

The stabilization system is composed of the ST-80, the alignment amplifier box, and the servo-loop amplifier box. These three units provide the missile with a space-fixed reference.

The guidance system contains the lateral and range computers. The two computers, in conjunction with their respective accelerometers located on the ST-80, provide the missile position with reference to the standard trajectory.

The program system contains the program device and the missile step switch and provides a time base for the missile.

The power system is composed of an 1800 VA inverter, two 28-volt self activating d-c batteries, and one 60-volt d-c power supply. The 1800 VA inverter provides 400-cps, 3-phase, 115-volt output for missile components that require highly regulated power, such as the program device. One 28-volt battery is utilized to drive the inverter; the other battery is used for general networks. The 60-volt d-c power supply is utilized for all command circuits, such as the command potentiometers.

The REDSTONE control system utilizes a control computer, a relay box, four actuators on the thrust unit, four actuators on the body unit, and four sets of spatial gaseous nitrogen jets, also located on the body unit.

Attitude control is maintained at all times. Lateral guidance is active during Phases I and IV. Range guidance is active only during Phase IV, but it computes engine cutoff during Phase I. Programing is maintained throughout flight in order to assure proper missile attitude. The control computer accepts these signals and mixes them in proper proportion for distribution to the various control surfaces. These units are interconnected as shown on the block diagram in Figure IV-2.

PROGRAM DEVICE

The program device is a three-channel magnetic tape recorder that uses 16-millimeter, 450-second Mylar tape. The information is recorded on the tape in the form of 2-kc pulses of 50-millisecond duration. Channel I provides the pulses for pitch programing which determine the missile trajectory. Channel II contains the pulses for the missile step switch, which is used for timing the flight sequence. Channel II also contains the dive program, which is activated at "Q", and is used for terminal guidance. Channel III is used for telemetry.

Tape motion is provided by a 115-volt, 400-cps, 3-phase motor which is coupled to 16-millimeter sockets through a mechanical differential. The tape is held under tension by a 28-volt d-c motor. This motor also compensates for the difference in speed as the tape moves from one reel to the other.

IV-3

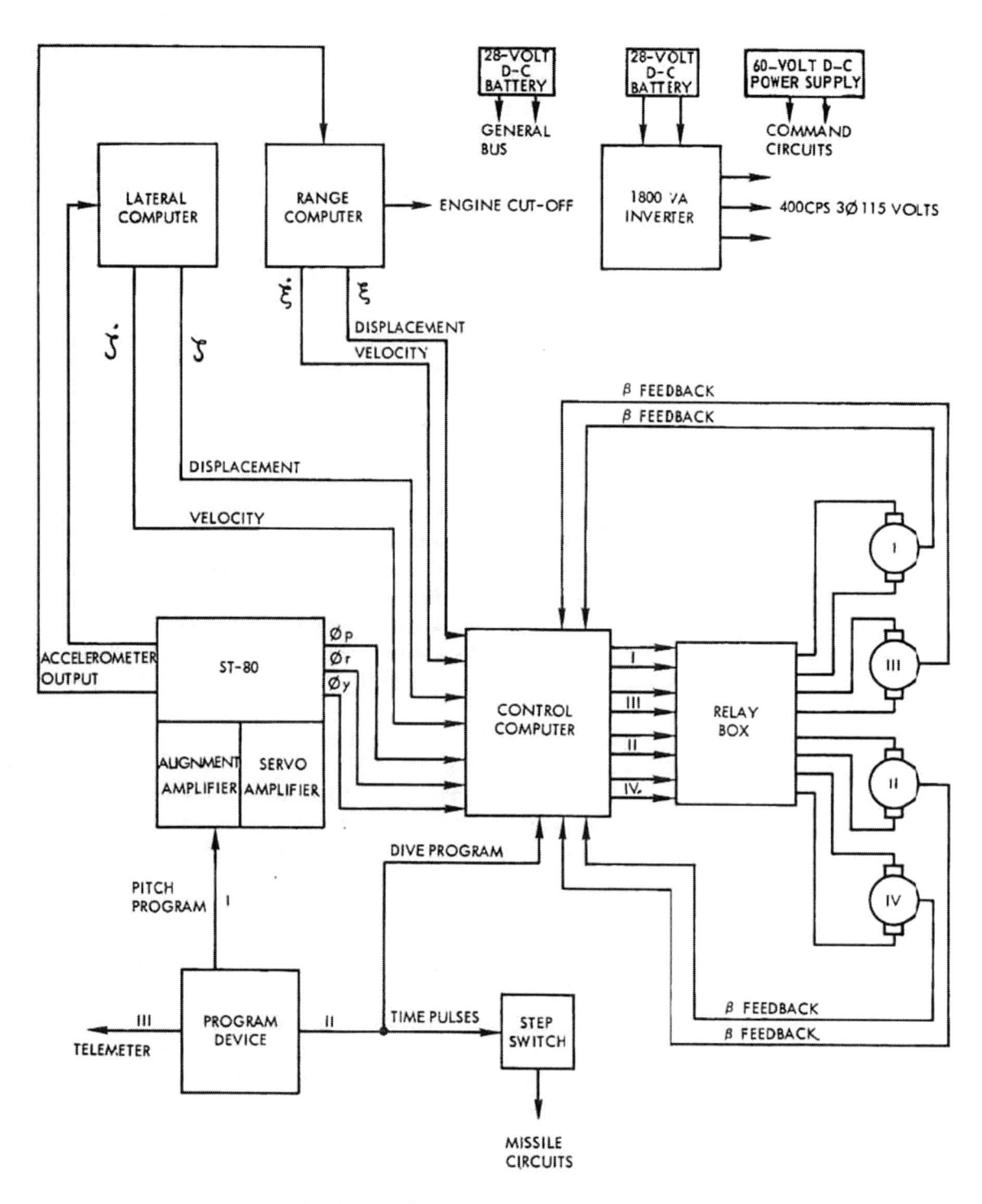

Figure IV-2 – Guidance and Control Block Diagram

The first time the program device is operated, a brake spring on the reel motor builds up tension which is released only upon a power failure. This increases reliability and prevents tape spillage during transit and operation. Transistor amplifiers, one for each channel, amplify the pulses from the tape and apply these pulses to output relays.

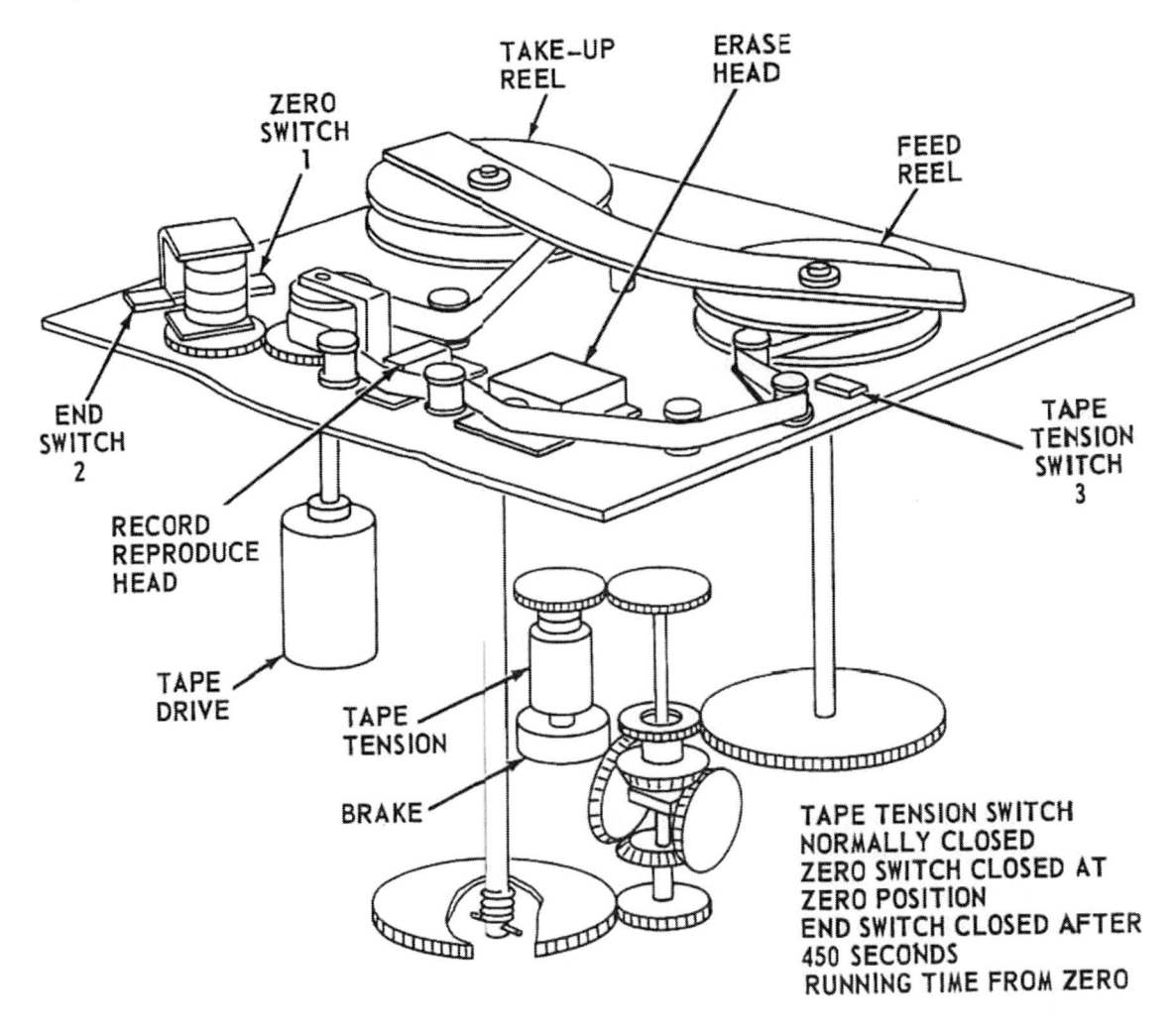

Figure IV-3 – REDSTONE 6700 Program Device

STABILIZATION SYSTEM

The accuracy of an inertially guided missile, such as the REDSTONE, is based primarily on the ability of the stabilization system to provide and maintain, throughout the flight, a space-fixed reference. Since the guidance and control error-sensing devices are mounted on this space-fixed reference, the need for stringent requirements to maintain this reference is obvious.

The stabilization system may be divided into four units: the ST-80 stabilized platform, the servo-loop amplifier box, the alignment amplifier box, and the stabilizer control panel. The following breakdown of each major component of the stabilization system indicates the location of minor components:

ST-80 Stabilized Platform

    3 stabilizing gyros (x, y, and z), type AB-9.
    3 servo motors (pitch, roll, and yaw).
    2 air bearing pendulums (x and z).
    2 integrating accelerometer assemblies (range and lateral), type AB-5.
    3 command potentiometers (pitch, roll, and yaw).
    Program transmission unit.
    Internal gimballing.
    Caging assembly.
    Azimuth inductive pickup.
    Mainshaft – including centerpiece.

Servo-loop Amplifier Box

    3 servo-loop amplifiers (stabilizing).
    3 servo-loop amplifiers (accelerometers).
    Various relays and transformers.

Alignment Amplifier Box

    3 torquer amplifiers (pitch, roll, and yaw).
    2 pendulum bias circuits.
    3 earth rotation bias circuits.

Stabilizer Control Panel

    Remote pendulum and earth-rotation bias adjustments.
    System energization circuits.
    System checkout meters and controls.
    Miscellaneous switches and indicators.

DESCRIPTION OF COMPONENTS

Stabilizing Gyros (x, y, and z gyros, type AB-9)

The rotor motor is a synchronous hysteresis type that uses the mass of the gyro rotor as the rotor of the motor. This motor rotates at 24,000 rpm and utilizes an air bearing about the precession axis of the gyro. An inner cylinder separated by air from the outer cylinder precesses with the gyro. Gyro precession is caused by the application of torque to the platform and gyro. An inductive pickup is used to provide an electrical output signal which denotes that the gyro has precessed in one direction or the other.

Accelerometers (range and lateral, type AB-5)

The rotor motor is a hysteresis synchronous type operating at 12,000 rpm. It utilizes an air bearing about the input, or sensitive, axis. An unbalanced weight or mass is mounted on the inner cylinder. When the platform experiences an acceleration or deceleration, the unbalanced mass applies a torque to the gyro and causes it to precess. A synchro transmitter attached to an output gear on the precession axis transmits the information to the guidance computers.

Servo Motors (pitch, roll, and yaw, 2-phase induction type)

Each of these motors has a fixed phase, while the other phase is provided from the outputs of the respective servo amplifiers. The motors apply torque between the stabilized portion of platform and the main shaft or missile.

Servo Amplifiers (pitch, roll, and yaw)

These amplifiers are practically identical. They receive a signal input from the inductive pickups of the stabilizing gyros, amplify it, provide the necessary phase shift, and feed the signal to the respective servo motors.

Air Bearing Pendulums (X and Z)

These pendulums are used to level the platform about the X and Z axes. The pendulums each use a slug which rides on an air bearing. When the platform is perfectly level, the slugs are positioned equidistant from either end of the pendulums, and the electrical output is zero. An off-level condition causes the slugs to move to one end or the other; the movement produces an electrical output, the phase of which denotes the direction in which the platform is off-level.

Azimuth Inductive Pickup

This pickup is used to determine when the stabilized portion of the platform (carrier ring) is perpendicular with respect to the mainshaft. It is a Dual-C type transformer, with a movable portion of the core (pole piece) attached to the carrier ring. When the platform is perpendicular to the mainshaft, the electrical output is zero. Being off the mechanical zero in either direction produces an electrical output with a phase indicative of the direction.

Command Potentiometers

Pitch, roll, and yaw potentiometers are used to provide the attitude signal voltages. These potentiometers are so attached on the platform that body and wiper move with respect to one another when the missile's attitude changes with respect to the stabilized platform.

Program Transmission Unit

This unit contains a program motor, a program solenoid, associated gearing, and a resolver. It is used to step the pitch potentiometer at a preset rate and preset distance to introduce a false error signal into the control computer. This unit receives from the program device the pulses necessary to operate the program solenoid. When the pitch command potentiometer is rotated, the rotor of the resolver rotates also. The resolver performs the interchange of signals from the roll and yaw gyro inductive pickups to the roll and yaw servo amplifiers as the missile pitches from vertical to horizontal and back to vertical.

Caging Assembly

This device is a motorized assembly which mechanically positions and locks or unlocks the carrier ring when power is to be removed from the stabilizer system, or after power is applied to the system.

Internal Gimbal

The internal gimbal is a system of gearing and bearings located partially on the carrier ring and partially in the mainshaft. This system provides freedom of motion of the platform in all axes.

Pendulum Bias Circuit

This circuit provides an electrical voltage to equal and cancel an output from the pendulum if the electrical and mechanical null are not in coincidence.

Torquer Amplifier

The X, Y, and Z amplifiers receive their input from the pendulums or azimuth inductive pickup, amplify it, and apply it to the torquer coils on their respective gyros.

Earth-Rotation Bias

While the platform is operating on the ground it is fixed to a point in space. As the earth rotates, there is a relative movement between the earth and the platform, thus causing the pendulums and inductive pickup to produce an output. The bias will torque the platform and cause it to precess at the earth's rate to keep it in coincidence, and to keep the outputs of pendulum and azimuth inductive pickup at zero.

FUNCTION OF THE STABILIZATION SYSTEM

The state of the art at the present time requires that the missile be mechanically attached to the platform through the gimbal system. Although the friction in the gimbal system is small, it is not negligible. Because of this friction, a torque is applied to the platform when the missile moves about its center of gravity or about its longitudinal axis. The stabilization system has the function of applying an equal and opposite torque to the platform to counteract the torque being applied by the missile movements, so that the platform does not deviate from its original setting. It is important to note that the stabilization system does not try to keep the missile from deviating in attitude. It merely keeps the platform from being dragged along with the missile. The relative movement between the missile and the stabilized platform is then measured electrically, and this output is called the phi signal.

The integrating accelerometers that are mounted on the platform are used to sense acceleration and deceleration along a particular coordinate. The range accelerometer senses acceleration and deceleration along the range coordinate, and the lateral accelerometer senses acceleration and deceleration perpendicular to the trajectory and in the horizontal plane. It is extremely important that these accelerometers do not change their measuring directions.

The ST-80 stabilized platform utilizes three air bearing powered gyroscopes. The principles of gyroscopic phenomena include that of rigidity and precession. The rules governing their action are:

- Rigidity — a body rotating steadily about an axis will tend to resist changes in the direction of the axis.
- Precession — that property of a spinning mass in which the plane of the spinning mass turns or tilts in the direction of applied torque.

A rotor of a typical gyroscope is shown in Figure IV-5. The arrow on the rim indicates the direction of rotation or spin. If a torque were applied to the gyro at any point on the shaft of the rotor, it would be effectively the same as if the torque were applied to a point on the rim in the same plane. The bottom half of the illustration shows torque effectively being applied to various points on the rim. If the rotor were spinning and point A' were the point of effective torque, the rotor would not move in the plane of applied torque, but rather at a point 90 degrees away from A' in the direction of rotation (point A") and in the same direction of torque. This phenomenon is true only if the rotor is spinning.

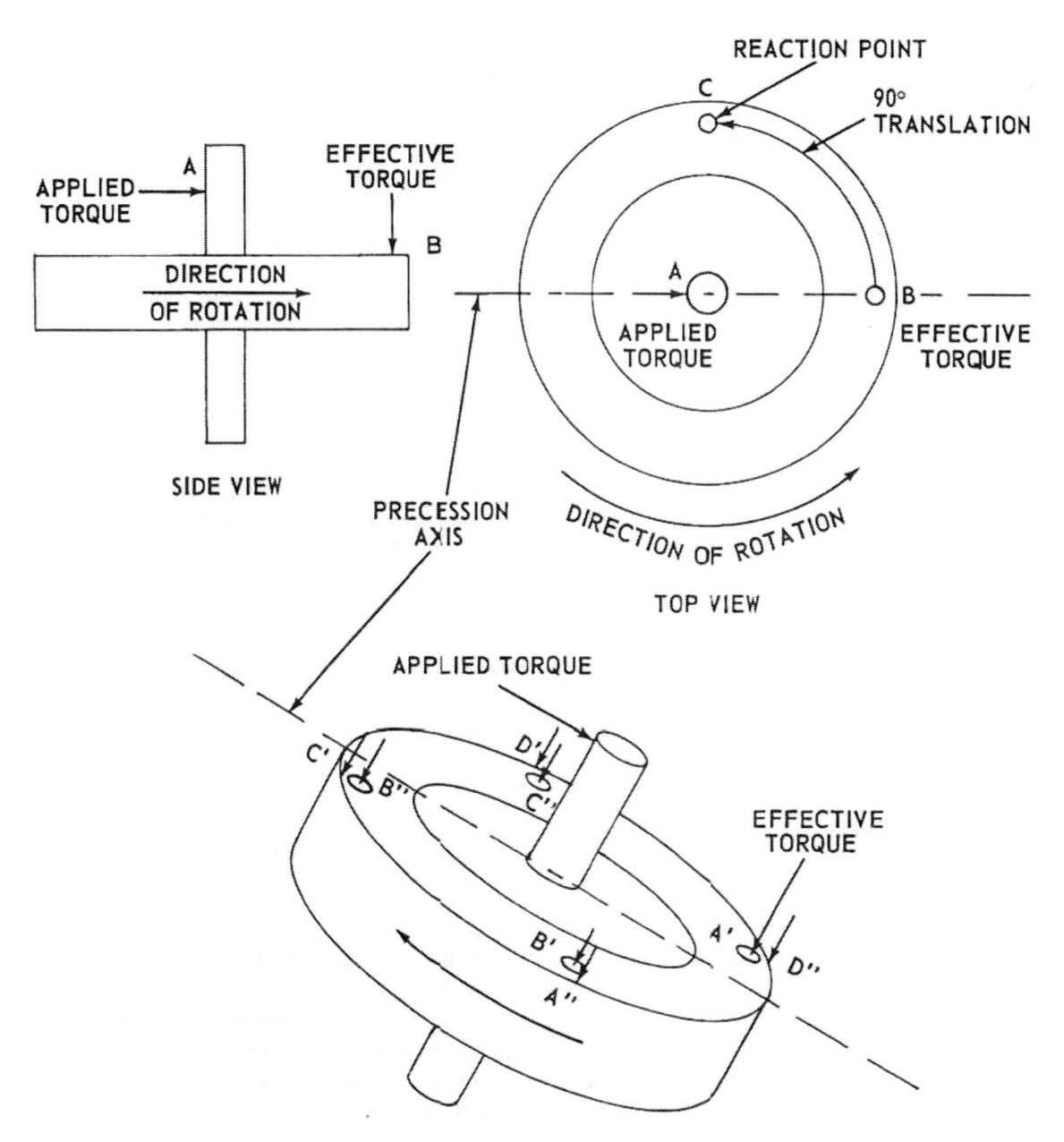

Figure IV-4 — Gyroscopic Spin and Precession

IV-10

The air bearing principle relies on a steady flow of air to support or lubricate the bearing surface. The lightest oil available has enough viscosity to cause an erroneous output signal from the gyro inductive pickup. Figure IV-6 is an idealized drawing depicting the method used to separate one bearing surface from another by use of air flow. The use of air bearings on the precessional axis of the stabilizing gyros reduces errors to a minimum.

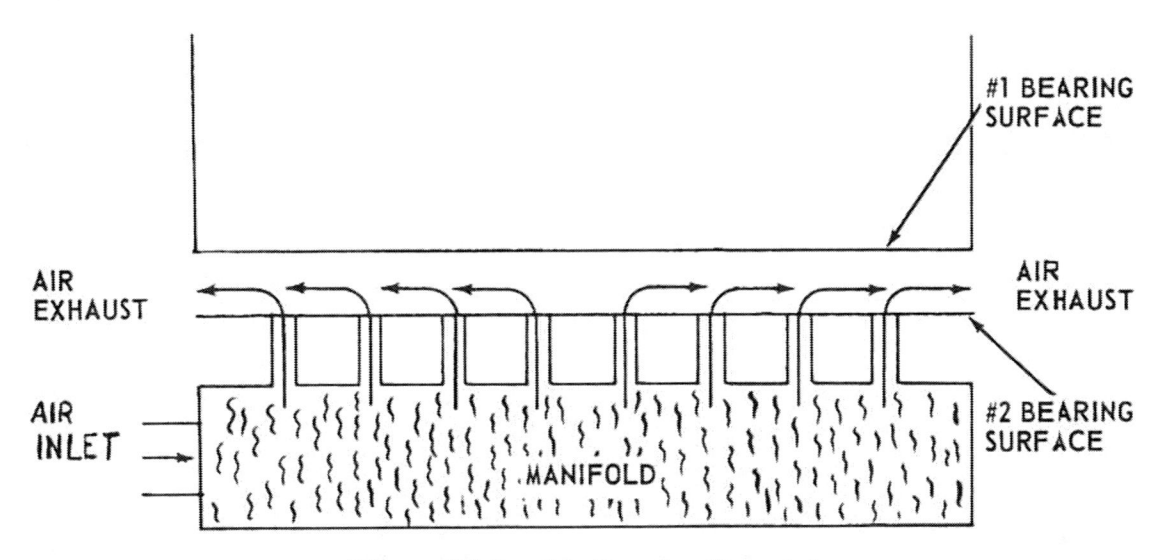

Figure IV-5 – Air Bearing Principle

The stabilization system's first problem is to level the ST-80 and to orient or aim it in the firing direction. This leveling and aiming is accomplished by means of the alignment circuitry and the stabilization circuitry. In leveling the platform to the local horizon, the x and z pendulums are used. Considering the pitch, or z axis, leveling, the z pendulum slug will be off level, thus an electrical output voltage with a phase indicative of the direction off level will be produced. This voltage will be fed into the torquer amplifier where it is amplified and fed back to the torquer coils mounted on the z stabilizing gyro. The torquer coils apply a torque to the gyro, which reacts by attempting to precess about an axis perpendicular to its normal precession axis. The platform responds by beginning to move in a direction of gyro precession. The gimbal friction then causes the gyro and inner cylinder to precess about the normal precession axis, and the inductive pickup is displaced. This produces an electrical output which is fed into the servo amplifier, where it is amplified and fed back to the pitch servo motor. The pitch servo motor drives against the pitch gear which is mounted effectively on the mainshaft. This removes the friction and allows the platform to rotate about the mainshaft until it has reached the point where the z pendulum becomes level and its output is reduced to zero. The same procedure is used for the leveling of the x pendulum.

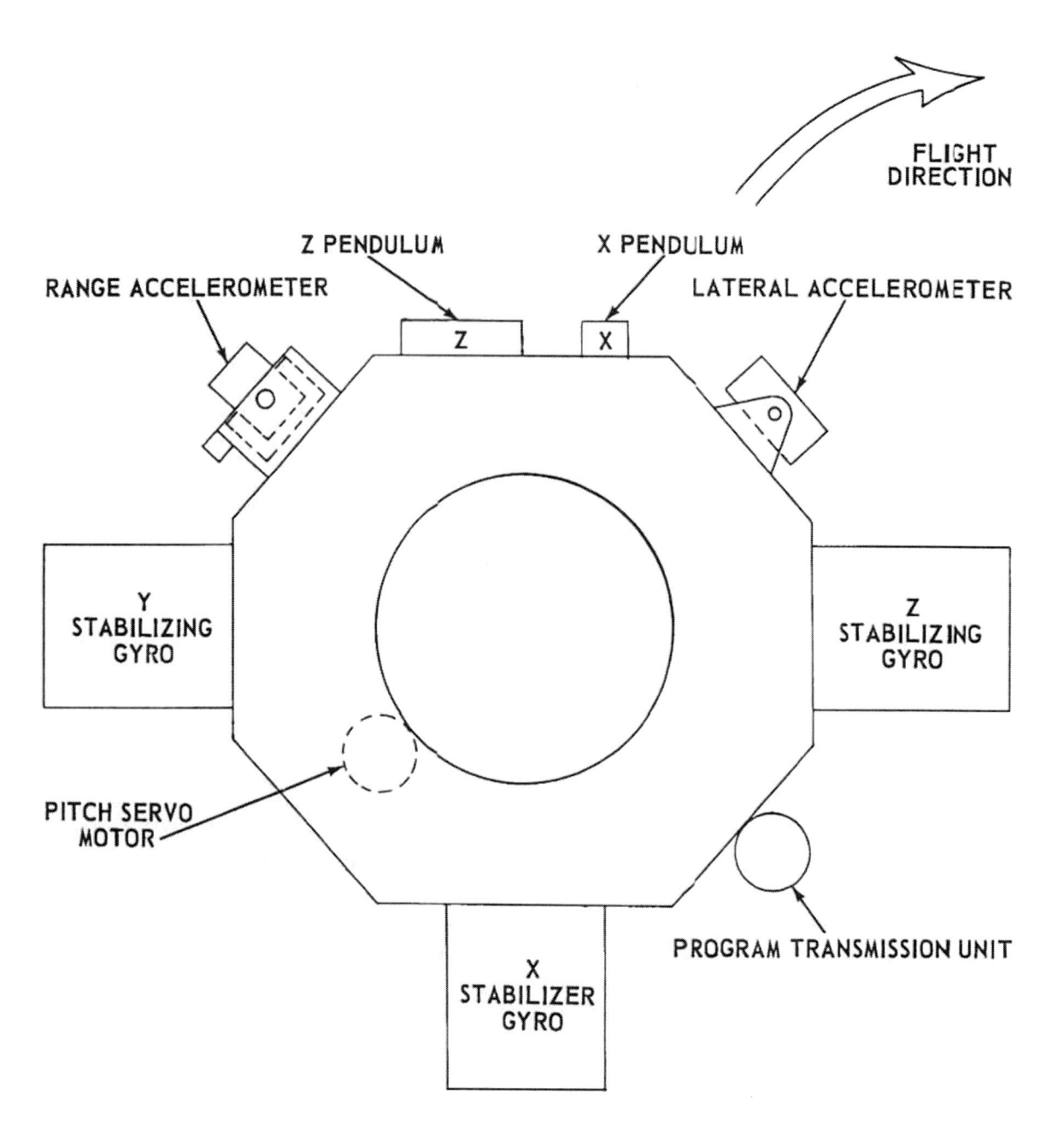

Figure IV-6 — Stability of the Platform Ring

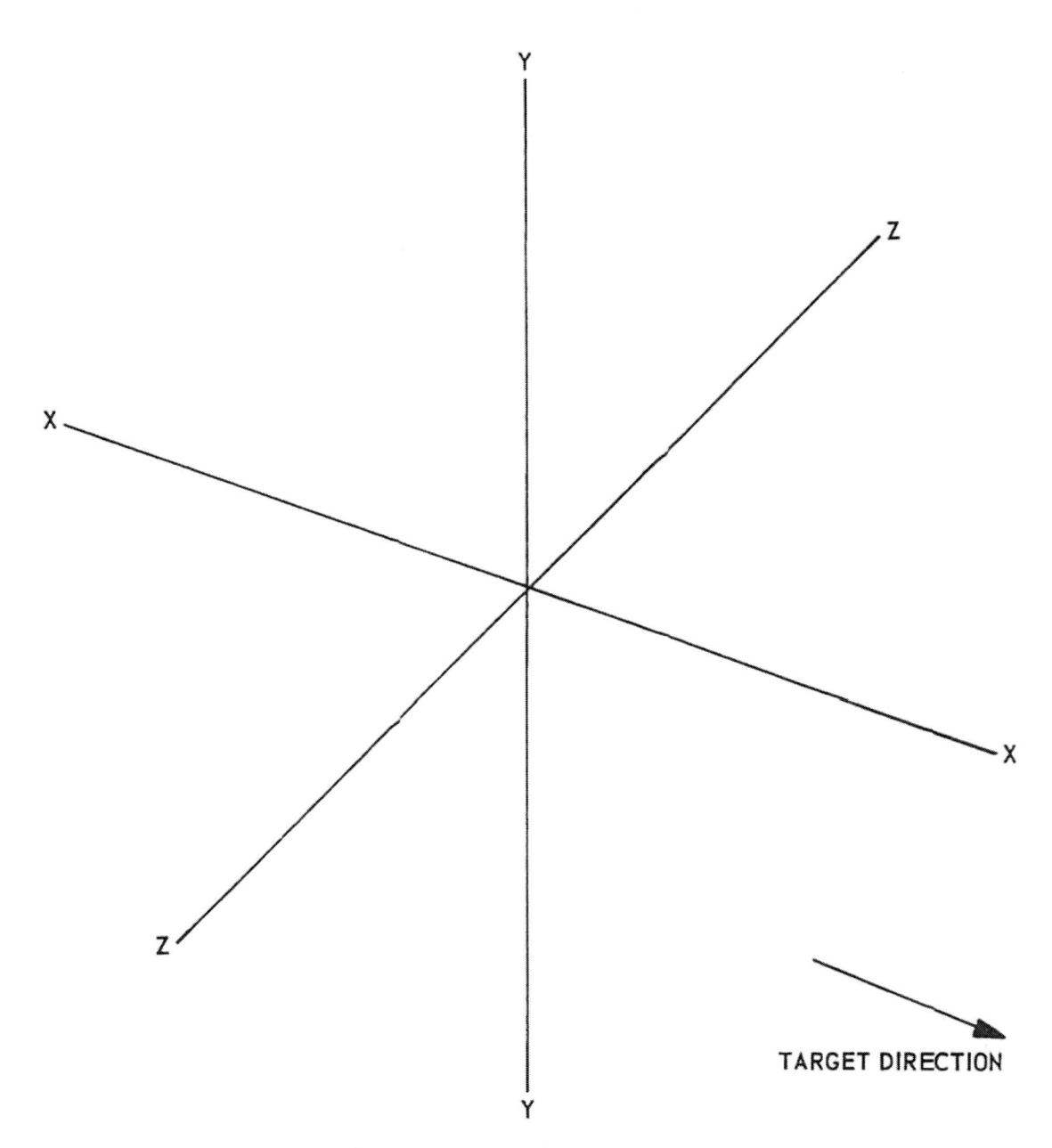

Figure IV-7 — Platform Coordinates

The azimuth inductive pickup senses when the mainshaft of the platform is not perpendicular to the carrier ring. The missile is oriented to the aiming azimuth by means of the porro prism and theodolite. The platform must then align itself to the missile or mainshaft. (The mainshaft of the ST-80 is attached to the missile.) The procedure is the same as that for the x and z channels with the exception of the sensing device, that is, the pendulum versus the azimuth inductive pickup.

In all three loops, the function is to continue to operate until the pendulums are level and the azimuth inductive pickup is nulled. These loops are then termed "null-seeking loops." However, due to the rotation of the earth, complications arise as long as the missile is on the launcher.

The platform is considered to be correctly aligned and oriented only when the pendulums and azimuth inductive pickup have zero output. Because of the rotation of the earth, the pendulums and azimuth inductive pickup will never really be zeroed out, for as soon as the servo motor stops turning, the rotation of the earth upsets the zero or nulled condition. Therefore, the platform could never be considered correctly aligned. Because of this situation, earth-rotation bias circuits have been incorporated into the alignment loops. Voltages are preset from the ground equipment and the value is predetermined and influenced by the geographic location of the firing site. This voltage causes the torquer coils to be constantly torquing the three gyros by an amount that will keep the platform perfectly leveled and oriented, thus keeping the pendulum and azimuth inductive pickup outputs at zero. This alignment circuitry will function until liftoff and will keep the platform earth-fixed. At liftoff the alignment circuits are de-activated; the platform then becomes space-fixed. The stabilization loops will then maintain the platform in its space-fixed position.

Basically, all three stabilization loops function identically. Roll and yaw, however, are modified by the use of a resolver to compensate for the interchange of the torquing planes of the two servo motors as the missile is pitched over from a vertical position to a horizontal position.

If a gust of wind causes a vertically rising missile to start pitching over, a torque is applied to the platform. Since a pitching motion is about the z axis of the platform, the z gyro will sense this torque because of gimbal friction about the Z axis. The reaction of the gyro would be to precess. As soon as the gyro moves the slightest distance, the inductive pickup will no longer be nulled, and consequently, an output voltage will be produced which will be of a particular phase indicative of the direction of the torque being applied to the platform. The output is fed to the servo amplifier, amplified, and fed back to the pitch servo motor. The pitch servo motor housing is mounted on the platform carrier ring, while the shaft and gearing drives a gear mounted on the mainshaft of the platform. Energizing the motor torques the gearing to cause the carrier ring to move in one direction or another about the mainshaft. The output of the servo amplifier causes the servo motor to apply a torque equal and opposite to that caused by missile movement. The rapidity with which the loop works allows the platform to move only a very small amount before correction is made. The small

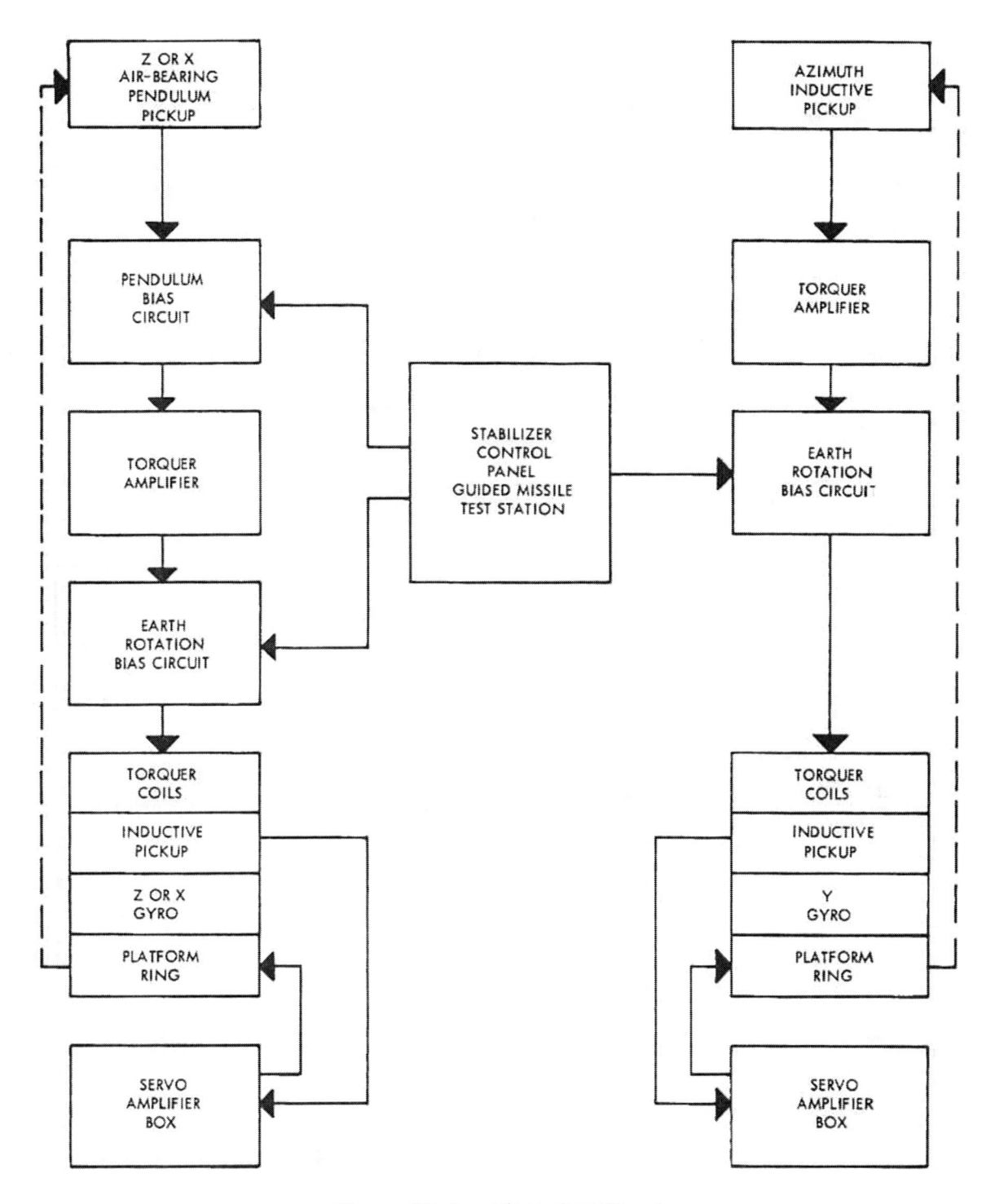

Figure IV-8 — Alignment Loops

IV-15

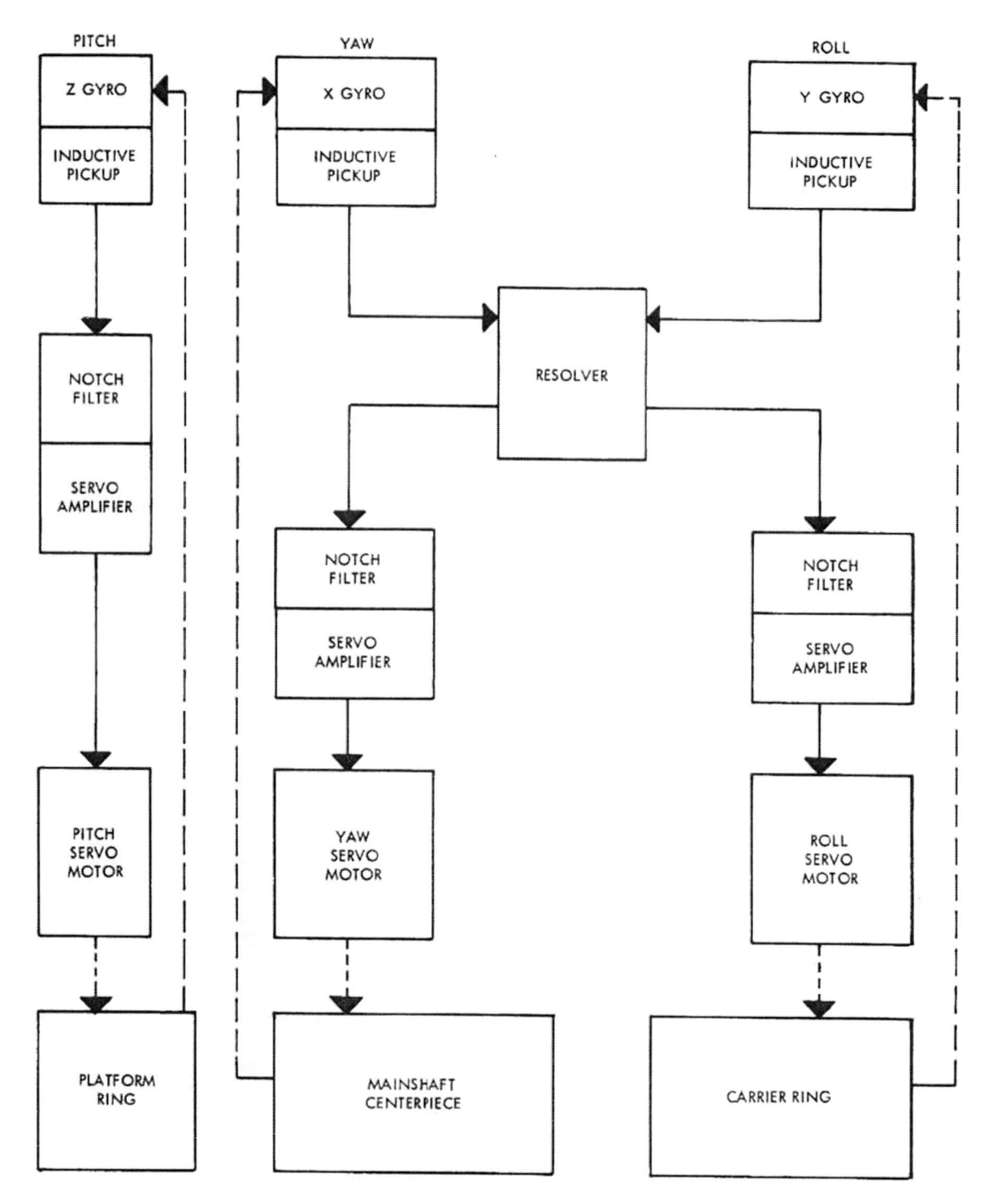

Figure IV-9 — Stabilization Loops

amount of error caused by the movement of the platform is corrected when the missile torque is removed. The servo motor will then drive to zero, or null out, the inductive pickup.

The yaw and roll channels work similarly, with the following exception: the outputs from the inductive pickups of the x and y gyros feed into a resolver. The output of the resolver is then fed to the servo amplifier. The function of the resolver is to apportion the input signals to the proper channel. The plane in which the roll and yaw servo motors apply countertorque changes in coincidence with the missile attitude about the pitch axis. Because a missile that is flying other than a true vertical or horizontal course will usually apply torque (to the platform) of such an angle as to cause at least two gyros to sense the torque, it is apparent that both servo motors would drive. The amount that each drives is dependent upon two factors: 1) the missile movement itself, as to whether it is a pure roll or yaw motion, or whether it is a complex motion involving both roll and yaw; and 2) the angle of the trajectory, or more accurately, the angle of rotor of the resolver. As an example, if the missile were flying at an angle of 45 degrees and it made a pure yaw motion, both the x and y gyros would sense the torque on the platform. Both gyros would produce an output to the resolver. However, due to the change of plane of the yaw and roll servo motors, the motor which we must drive to counteract this torque is the yaw servo motor only. Without the resolver, both would drive an equal amount and the resultant applied torque would not be exactly opposite to the applied torque and our reference would be lost. Because of the resolver action, however, the only output from the resolver will be that to the yaw servo amplifier and motor.

With the platform stabilized, the problem is to measure the deviation of the missile with respect to the reference. This is accomplished by using command potentiometers. With the body of a potentiometer attached to the mainshaft and the wiper attached to the carrier ring, any relative movement between the two will produce an output voltage. This voltage will be either a positive or negative voltage with respect to a reference voltage. The differential voltage is fed directly to the control computer, where it is used to correct the attitude of the missile. The wiper on the potentiometer is thereby repositioned and the output voltage is reduced to zero. All three attitude signals are derived identically in that the pitch, roll, and yaw attitude of the missile is measured by their respective potentiometers.

The use of the program transmission unit to provide the pitch programing necessary to tilt the missile into the ballistic trajectory works in conjunction with the pitch command potentiometer. The program transmission unit consists of the program motor, the program solenoid, the slip clutch, gearing, and the resolver. During flight, the program motor is turning constantly, but the shaft and gearing are attached to the motor by the slip clutch. Unless the solenoid is energized, the shaft is locked in position. The program device supplies the pulses necessary to energize the program solenoid. At a predetermined time of flight, it is desired to tilt the missile to coincide with the ballistic trajectory. Pulses are fed from the program device to the program solenoid, which is then energized and allows the shaft and gearing to turn. The length of the pulse and the paw and ratchet assembly will allow the shaft to turn the equivalent of 1 degree of missile pitch. The turning of the shaft will do two things: 1) it will rotate the resolver rotor 1 degree; 2) through a series of gears, it will slide the wiper of the pitch command potentiometer to a position which will produce voltage indicative of 1 degree of error. The control computer responds to this voltage by causing the missile to pitch over 1 degree and thereby repositions the body of the pitch command potentiometer to a null point. This process continues according to a predetermined trajectory.

The range accelerometer is the device which senses the acceleration and deceleration of the missile along the range coordinate. The epsilon angle is that angle formed by the range coordinate and the local horizon. Since the range coordinate is different for each change in range of the target, the epsilon angle is also different. In order to measure acceleration along the range coordinate accurately, the angle of the range acceleration must be adjusted in such a manner that the true measuring direction is coincident with the range coordinate. The use of the stabilizer control panel and a motorized epsilon setting assembly of the range accelerometer enables a precise setting of the correct angle.

Since the lateral accelerometer is affected only by deviations of the missile to the left and right of the target, it is mounted on the platform without available means of adjustment.

Fundamentally, both the range and lateral accelerometers function identically. An acceleration of the missile reacts on the unbalanced mass, or weight, attached to the inner cylinder that contains the gyro and applies a torque to the spinning mass about the air bearing. This will torque the gyro and will cause the entire accelerometer assembly (with the exception of the platform mounting) to precess. Attached to this assembly is an output gear to which is coupled a synchro transmitter and a servo motor. The synchro transmitter rotor turns, thus providing a rotating electric voltage to the guidance computer. This rotating voltage is indicative of the attained velocity. Stated another way, the angular rotation of the accelerometer about the precessional axis in a given period of time is a measure of the velocity. Thus it can be seen that the first integration from acceleration to velocity is accomplished in the accelerometer.

The function of the accelerometer servo loop is to remove friction in the precessional axis of both accelerometers. The accumulation of frictions resulting from slip rings, ball bearings, and gearing would result in erroneous velocity information. When the unbalanced weight is acted upon by acceleration or deceleration, it applies torque to the inner cylinder that contains the gyro. The gyro will then begin to precess. Since there is friction about the precession axis, some means must be provided to counteract this friction. On the end of the inner cylinder of the accelerometer is an inductive pickup which produces an electrical output as a result of the friction when the accelerometer begins to precess. This voltage is amplified and fed to the servo motor on the accelerometer assembly. The torque that it applies to the precession gear is exactly enough to cancel out the friction. The precession axis, therefore, is considered to be essentially frictionless, thereby providing the guidance computers with the correct value of velocity.

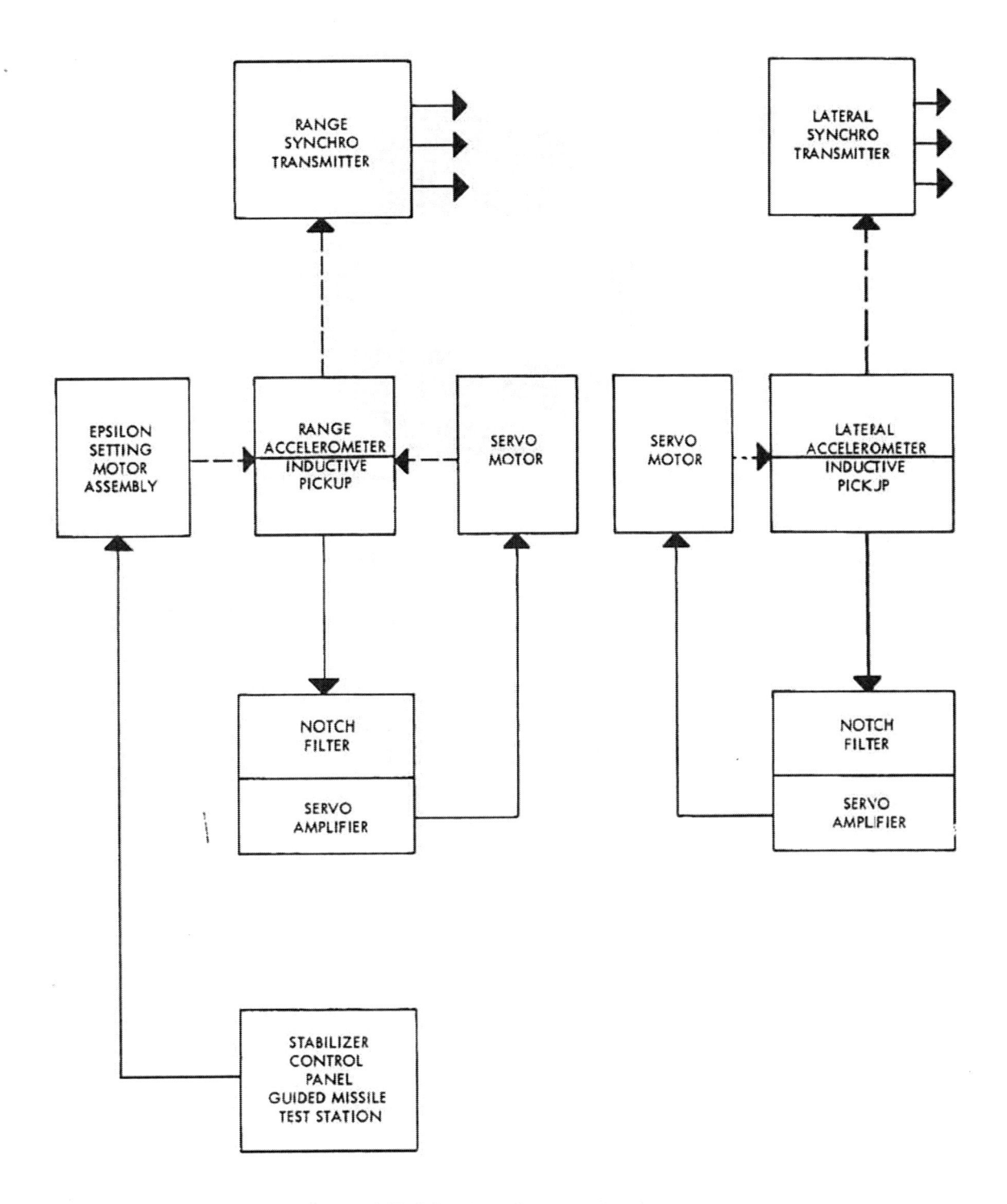

Figure IV-10 — Accelerometer Loops

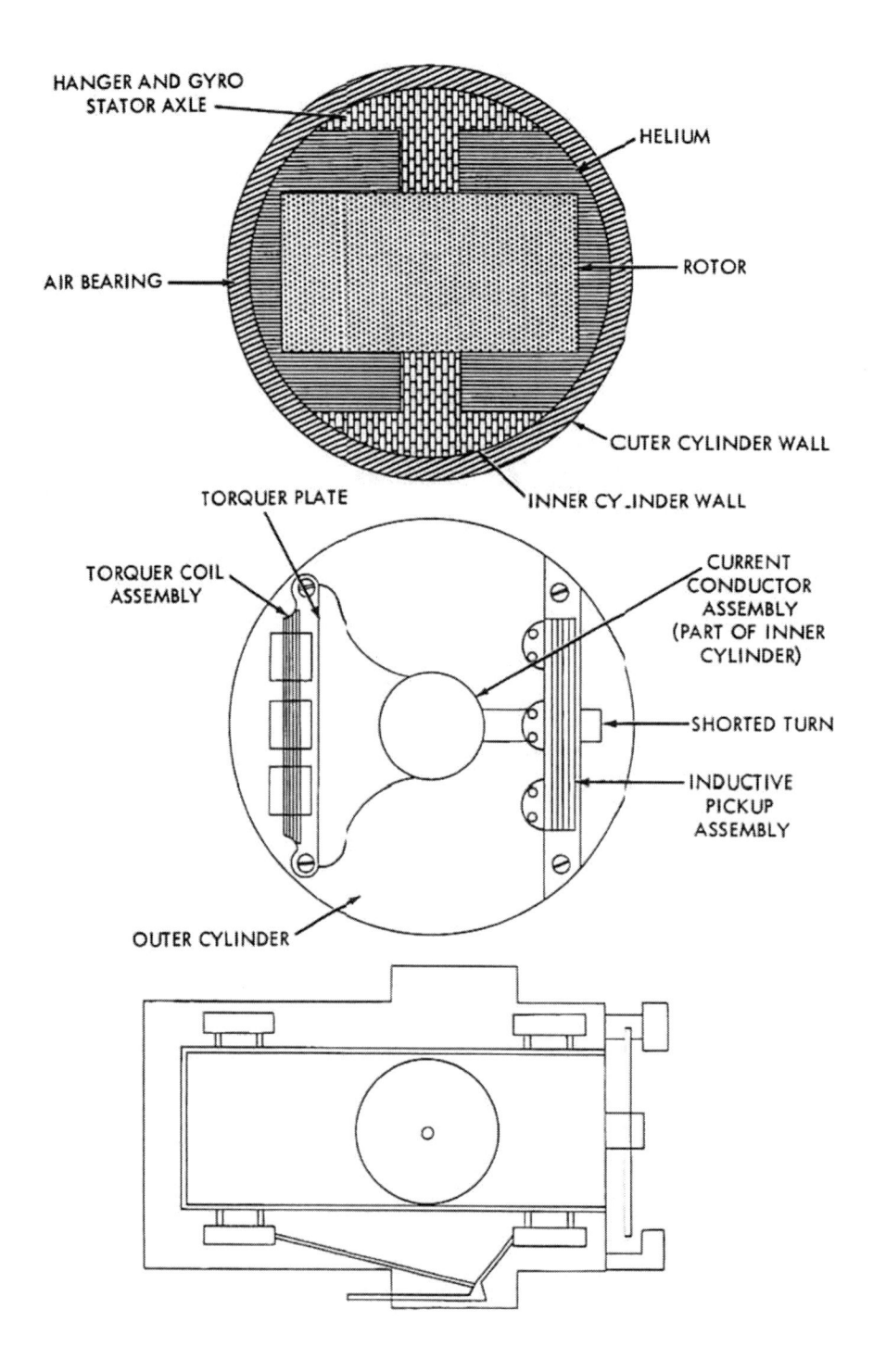

Figure IV-11 — AB-9 Gyro

GUIDANCE SYSTEM

The REDSTONE guidance system is divided into three sections: 1) lateral guidance computer, 2) the range guidance computer and 3) the cutoff computer.

The range and lateral guidance computers receive and store velocity information from the range and lateral accelerometers. In addition, the velocity information is integrated and the resulting displacement information is stored. At predetermined times during flight, this velocity and displacement information is fed into the control computer. The control computer then acts to cause the missile to alter its course to return to the predetermined trajectory. The cutoff computer utilizes the range velocity and displacement information to solve the cutoff equation. When this equation is solved (with some particular reservations), the thrust will be terminated.

Lateral Guidance

The problem of maintaining the missile's center of gravity on the trajectory is a function of the guidance computers. In the case of the lateral guidance computer, it serves to provide a corrective signal when the missile deviates to the left or right of the trajectory, from launch to cutoff plus 1.5 seconds. From cutoff plus 1.5 seconds to re-entry (Q), any deviation of the missile from the trajectory is sensed, and rather than providing corrective action immediately, the information is stored. At re-entry, this information is fed into the control computer, and the missile begins to make the necessary corrective maneuver to return to the intended trajectory. This last phase is referred to as terminal guidance.

Deviations from the trajectory can be caused by atmospheric conditions, thrust decay, separation, and other factors. The deviations are sensed by the use of integrating accelerometers. The action of the accelerometer is to precess whenever the missile accelerates or decelerates in the measuring plane of the accelerometer. For this reason, the lateral accelerometer is mounted on the stabilized platform in such a position as to sense acceleration and deceleration either to the right or left of the trajectory.

When a vertically rising missile is shifted to the left of the trajectory, the lateral accelerometer precesses at a rate proportional to the acceleration and in a direction dependent upon the direction of acceleration. The output coupling device is a synchro transmitter; the rotor is rotated mechanically by the precession of the accelerometer. The rotating rotor of the synchro transmitter will thus cause the stator voltage to rotate electrically at a rate proportional to the velocity. This rotating electrical signal voltage sets up a rotating magnetic field in the stator of the control transformer. A voltage is thereby induced into the rotor of the control transformer and this voltage, in turn, is fed to a servo amplifier, where it is amplified and used to drive a servo motor. The servo motor drives a gear train which performs three functions: 1) repositions the rotor of the control transformer to zero out the input to the servo amplifier, 2) positions the wiper of the lateral velocity control potentiometer to a point such that the voltage appearing on the wiper is indicative of the missile velocity, and 3) drives the lead screws of the ball-and-disc integrator. The ball-and-disc integrator performs the second integration, that of velocity with respect to time, and the output drives a gear train which positions the wiper of the displacement potentiometer. The position of the wiper determines the output voltage, and this voltage is then indicative of the displacement from the trajectory. Thus, two voltages are fed to the control computer: 1) lateral velocity signal voltage and, 2) lateral displacement signal voltage. These voltages will continue to increase as long as the missile is accelerating. Once the missile attains a constant velocity, the lateral accelerometer will stop precessing, the servo motor will stop driving the gear train, and the wiper on the velocity potentiometer stops moving. The ball-and-disc integrator output is still rotating, however, the wiper on the displacement potentiometer continues to move, and the displacement voltage continues to increase.

In Phase I (launch to cutoff) of the flight, however, the control computer accepts these input voltages and produces the necessary driving voltages to drive the air rudders in a direction which will return the missile to the intended trajectory. When the rudders move, the missile will decelerate laterally, the lateral motion of the missile will stop, and it will accelerate back toward the intended trajectory. As this happens, the lateral accelerometer will sense the deceleration and acceleration and will precess in the opposite direction. As it does, the servo system in the lateral guidance computer turns in the opposite direction, thus moving the wiper back toward the zero voltage point and repositioning the lead screws of the ball-and-disc integrator so that the displacement potentiometer wiper is also returned to zero. At this time, the missile will be back on the intended trajectory and the guidance system will become static until there is another disturbance.

During Phases II and III, the accumulated errors in velocity and displacement are stored on the potentiometers rather than being fed to the control computer. At reentry (Q), the potentiometers are zeroed as in Phase I.

Range Guidance

The primary purpose of the range guidance system is to assure that the missile does not overshoot or undershoot the target. Included in the system is the range accelerometer, the range computer, and the cutoff computer.

The range accelerometer is mounted on the stabilized platform; the range and cutoff computers are mounted on a chassis in the range computer box.

The information relative to this system is the velocity and displacement of the missile along the range coordinate. Therefore, the accelerometer is mounted on the stabilized platform at an angle such that the true measuring direction is coincident with the range coordinate. The angle subtended by the range coordinate and the local horizon is referred to as the epsilon angle.

The problem of putting the missile into a ballistic trajectory so that its point of impact can be determined involves some special consideration that must be provided for in the range computer. Preset information is inserted into the range computer's velocity and displacement potentiometers.

Since the missile launching site is physically ahead of the theoretical launch site (that site which would be used in conjunction with the reference trajectory to place a missile on the selected target) and the REDSTONE Missile starts from zero velocity (rather than a fixed specific velocity), a negative velocity and a positive displacement value are preset into the range computer and cutoff computer velocity and displacement potentiometers, respectively. In addition, preset values which allow for the effects of thrust decay, separation of the body and thrust unit, and air drag from cutoff to re-entry are inserted.

As the missile leaves the launcher, the range accelerometer begins to precess. As it does, the range servo motor begins to reposition the velocity and displacement potentiometers toward zero. When cutoff occurs, the preset velocity and displacement is considered to be cancelled out (the missile has reached the velocity and displacement of the theoretical missile). The only values remaining would be those of the effects of thrust decay, separation, and air drag.

From cutoff to re-entry, the change in range velocity is relatively small (and if conditions were exactly as assumed for in the presetting, the range computer would be zeroed at re-entry). At re-entry, any range information that remains on the range velocity and displacement potentiometers would be fed into the control computer, where corrective action would be initiated to zero out these errors.

Cutoff Computer

The function of the cutoff computer is to compute the proper time for thrust termination (engine cutoff) and to initiate the action. The computer is a magnetic amplifier type; preset negative velocity and positive displacement information is supplied to the computer from the cutoff velocity and displacement potentiometers. In addition, bias voltages representing calculated time from cutoff to re-entry and a cutoff constant which allows for the effects of thrust decay, separation, and air drag are fed into the computer. The preset velocity and displacement information must be zeroed out in the cutoff computer. Since the velocity and displacement are continually changing as the missile is ascending, the cutoff equation is solved several times. In order to prevent premature cutoff, a cam-operated switch which is in series with the cutoff relay in the cutoff computer must be closed. The cam switch is operated when the missile has reached within -55 to +60 meters per second of the desired velocity. The first time the cutoff equation is solved after the cam switch is operated (and providing the X+96 seconds relay is closed), the cutoff relay (in the control distribution) will be energized, thus initiating engine cutoff. After cutoff, the cutoff computer serves no function.

CONTROL SYSTEM

The REDSTONE control system as previously mentioned, consists of the control computer, the relay box, and actuators. The control computer receives the seven guidance and control signals and amplifies and mixes these signals in the right proportion for distribution to the control surface. Phi signals (pitch, roll, and yaw) pass through R-C networks and into the respective preamplifiers, where they are mixed with any incoming guidance signals. Lateral guidance signals will be fed to the yaw preamplifier, and the range signals will be fed to the pitch preamplifiers. The outputs of the preamplifiers are fed into the main amplifiers. The output of the pitch preamplifier is fed into main amplifiers II and IV. The yaw preamplifier feeds main amplifiers I and III. All four main amplifiers are fed by the roll preamplifier.

All amplifiers in the computer are magnetic amplifiers. Through the use of the R-C networks, the incoming phi signal can be broken down into its two component parts. The current flowing into the preamplifier through the resistance portion of the network represents the magnitude of error. A second current flows into the preamplifier through the capacitance portion of the network, which represents the rate of change of the error signal.

The output of the preamplifier will be a result of not only how much error has occurred but also the rate at which it occurred. The incoming velocity signals from the guidance computers go directly to the preamplifiers. However, the displacement signals are placed across two potentiometers which are driven by the step motor. At "Q" the step motor is driven by the dive program from the program device and, in turn, drives the wipers of the potentiometers from the zero position to the maximum position. This enables the displacement errors that have occurred during flight to be corrected for gradually instead of inserting the entire error into the system at once.

Also contained within the control computer are various relays. The purpose of these relays is to switch circuit values in order to change the gain and response of the control system to compensate for changing missile velocity, configuration, and aerodynamic conditions. The output from the four main amplifiers is fed to the relay box. The relay box, as its name implies, consists of relays which convert the small-amperage signal coming from the main amplifier of the control computer into the high amperage necessary to drive the actuator motors. The incoming signal energizes a polarized relay which, in turn, energizes a heavy-duty relay that connects the 28-volt battery to the actuator motor. As the actuators drive the vanes outward, a potentiometer (beta potentiometer) within the actuators feeds vane position information back into the main amplifier in the control computer, where it is compared to the original error. Since there are four channels, the relay box contains four polarized relays and four heavy-duty relays (one polarized and one heavy-duty relay for each of the four channels). The output of the relay box is 28 volts dc from the general network battery connected to the actuators through the heavy-duty relays. The actuators consist of a 28-volt d-c motor, a gear train, an output member, two potentiometers, and four microswitches. The gear train gives a reduction of approximately 600 to 1. The two potentiometers provide vane position information. The beta potentiometer feeds its information back to the control computer, and the measuring (M) potentiometer feeds information to the telemetry circuits during flight and to the ground checkout equipment during checkout.

POWER

There are two phases of operation of the REDSTONE system power sources. The first phase involves the use of ground support equipment, in addition to a portion of the missile power equipment, to supply necessary voltages to check out and prepare the missile for launching. The second phase is initiated by a step in the countdown procedure called power transfer. During the second phase, all power used on the missile is supplied by sources aboard the missile. The following list contains all the power sources used in the two phases:

Phase I

60-kw diesel generator — 208-volt/120-volt, 60-cps, 3-phase
2 aircraft energizers — 28-volt d-c
60-volt d-c power supply
2 inverters — 115-volt, 400-cps, 3-phase

Phase II

2 self activating batteries — 28-volt
60-volt d-c power supply
Inverter, 115-volt, 400-cps, 3-phase

In addition to the above equipment, the following panels, located in the guided missile test station and the power distribution trailer, are necessary for energizing, controlling, and monitoring these components:

Precision generator
Inverter control panel
Electric power panel
Power supply panel, 28-volt d-c

Phase I

The function of the ground power equipment is to supply the necessary voltages to be used for checkout, testing, and launch preparation.

The primary power source is the 60-kw diesel generator. This generator supplies 208-volt, 3-phase, 60-cps voltage to the a-c distributor box, the battery service trailer and the power distribution trailer. The generator is driven by a 6-cylinder Cummins diesel engine that supplies 160 hp at 2,500 rpm. Its normal operating speed is approximately 1,850 rpm. This engine drives a General Electric synchronous generator which provides a 208-volt, 3-phase, 60-cps output from a Y-connected transformer.

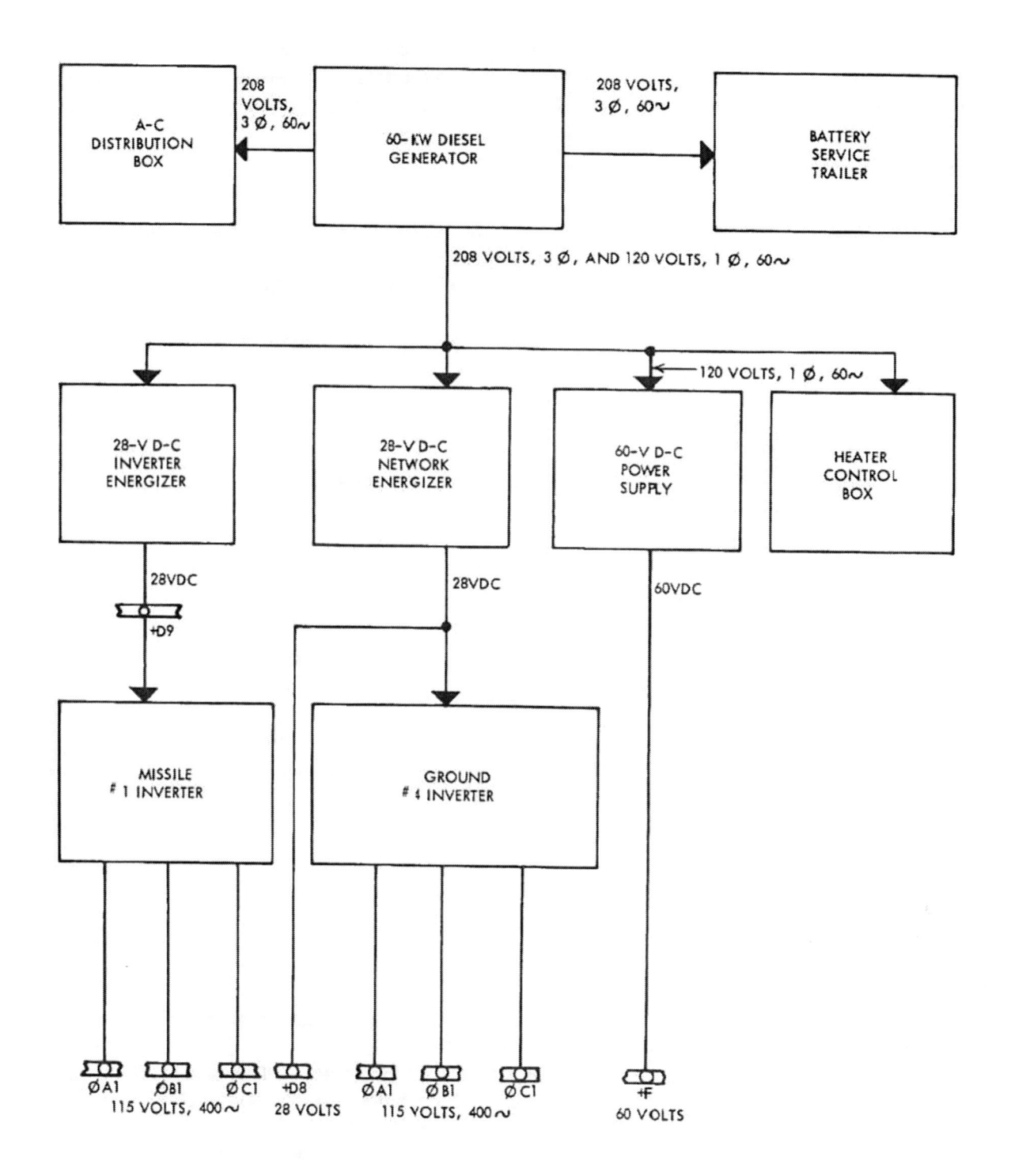

Figure IV-12 — Primary Power Source

MISSILE VOLTAGE DISTRIBUTION

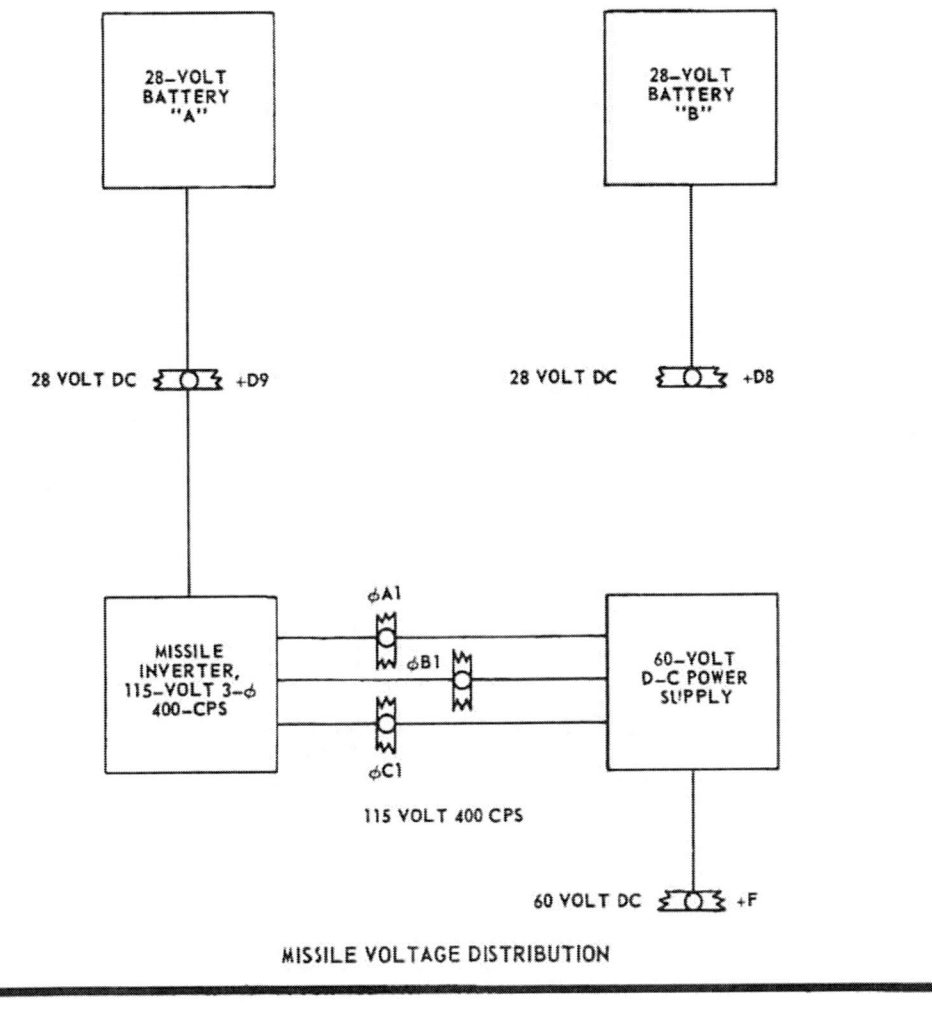

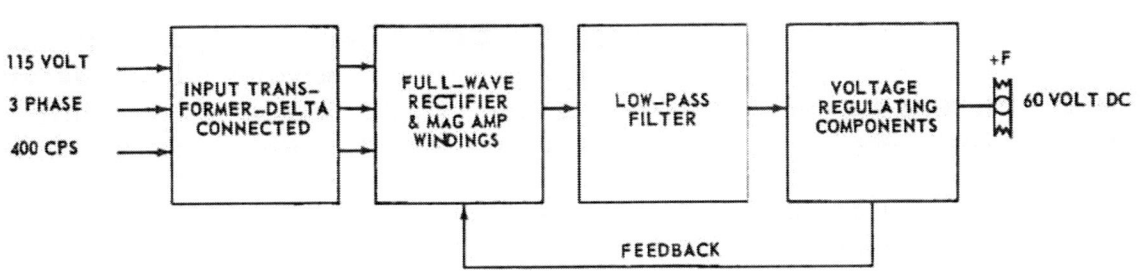

60-VOLT D-C MISSILE POWER SUPPLY

Figure IV-13 — Missile Power Components and Distribution

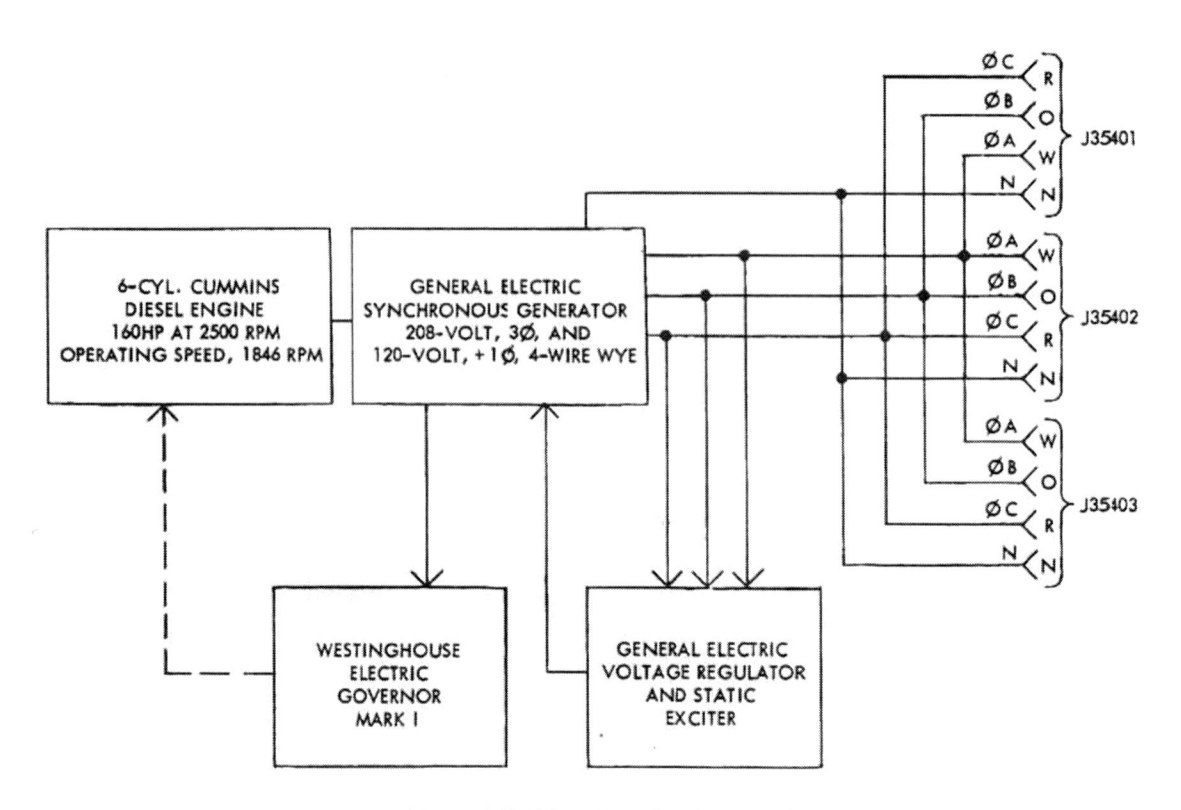

Figure IV-14 — 60-KW Generator

The 4-wire output, in conjunction with the 208 volts (phase-to-phase), also makes available 120 volts (phase-to-neutral). It employs both frequency and voltage regulation. The frequency regulation is accomplished by a Westinghouse Electric governor and the voltage regulation by a General Electric voltage regulator and static exciter.

The voltage supplied to the a-c distribution box is fed to various pieces of ground support equipment such as the alc and the LOX trailers, the hydrogen peroxide vehicle, and the guided missile test station.

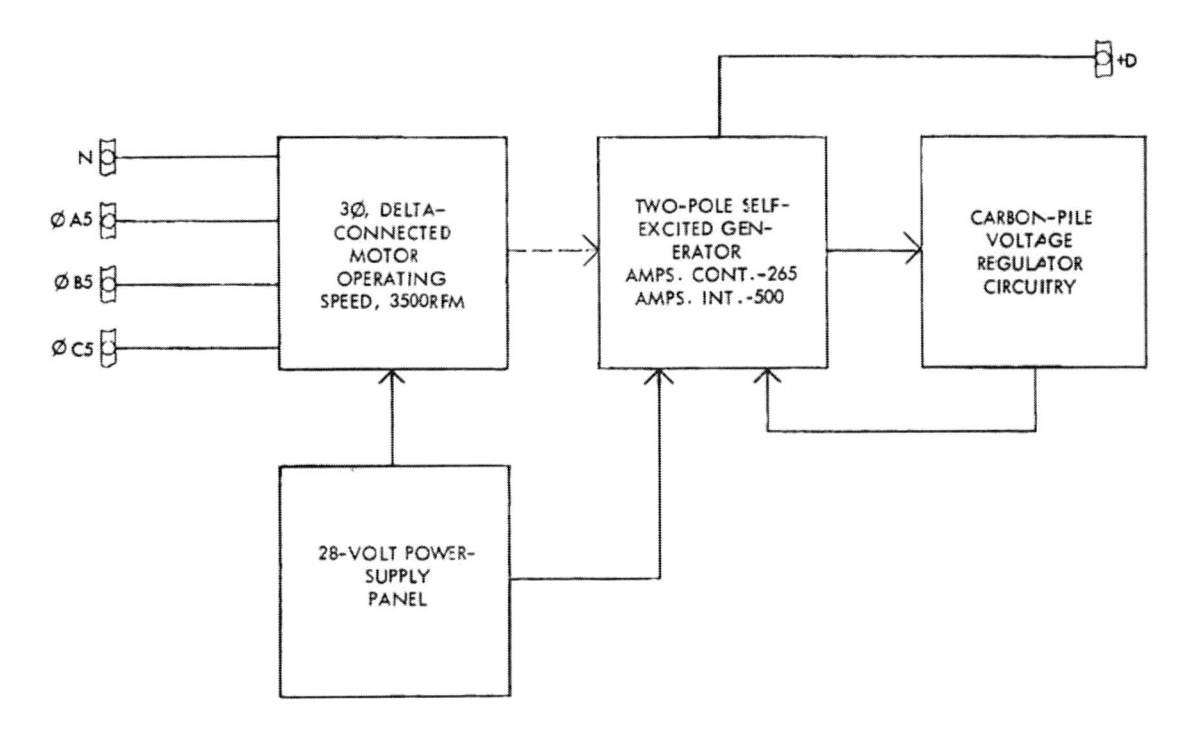

Figure IV-15 — 28-Volt Energizer

In the power distribution trailer, the 208-volt, 3-phase voltage is fed to the two aircraft energizers. These motor generators provide the system with 28 volts dc. They consist of a 3-phase delta-wound motor driving a generator which provides 28 volts dc at up to 265 amperes continuously. It utilizes a carbon pile regulator to provide voltage regulation. The controls for energizing, monitoring, and adjusting are located on the 28-volt d-c power supply panel. The outputs from the generator are fed to two busses. The network energizer is connected to the +D1 bus which feeds the +D8 bus in the missile. This bus is used primarily for operating relays and the rotary actuators. The other energizer provides voltage to the +D2 bus, which feeds into bus +D9 on the missile. This bus is used primarily for operating inverter 1 on the missile.

The 60-volt d-c power supply is used to provide voltage for the command potentiometers, the velocity and displacement potentiometers, and the beta feedback potentiometers. Its input is 120 volts, 60 cps, single phase and has an output of 54 to 66 volts dc at up to 2 amperes. It uses a magnetic amplifier type voltage regulator. Its output is fed to the +F bus (Figure IV-16).

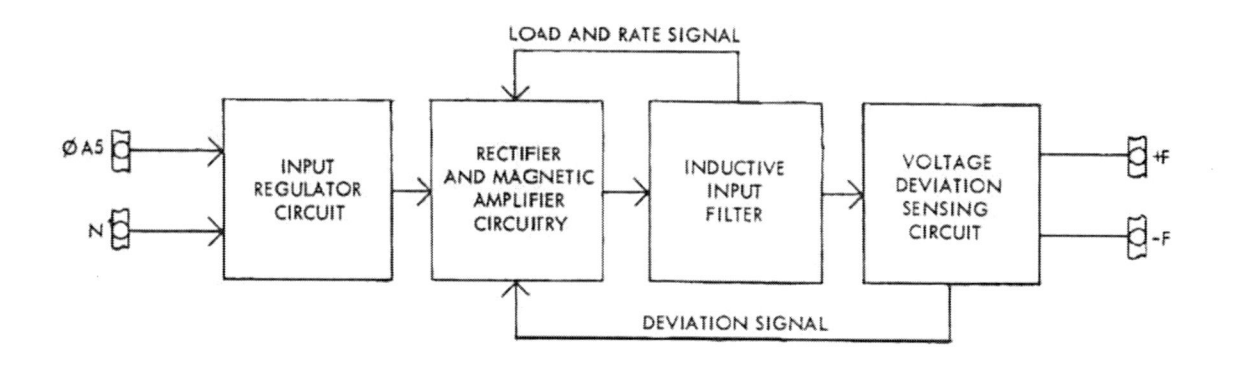

Figure IV-16 – 60-Volt Regulated Power Supply

Inverter 4 is used to provide 115-volt, 3-phase, 400-cps voltage to the ground equipment test fixture. The inverter consists of a 28-volt compound wound motor which drives a 3-phase, Y-connected generator. Included also is a regulator which contains both frequency- and voltage-regulator circuits. By the use of a frequency discriminator circuit, a change in frequency can be quickly detected and the current in the shunt winding of the motor increased or decreased to alter the speed of the motor, thus changing the frequency. A voltage discriminator circuit detects changes in output voltage and controls the current in the generator field, thus controlling the output voltage.

The electric power panel located in the guided missile test station contains meters to monitor the two 28-volt d-c busses. Also included on this panel is a meter which is used to monitor the 60-volt d-c command voltage supply and a switch which energizes a relay to apply the voltage to the +F bus.

Inverter 4 is used to provide 115-volt, 3-phase, 400-cps voltage to the ground equipment test fixture. The inverter consists of a 28-volt compound wound motor which drives a 3-phase, Y-connected generator. Included also is a regulator which contains both frequency- and voltage-regulator circuits. By the use of a frequency discriminator circuit, a change in frequency can be quickly detected and the current in the shunt winding of the motor increased or decreased to alter the speed of the motor, thus changing the frequency. A voltage discriminator circuit detects changes in output voltage and controls the current in the generator field, thus controlling the output voltage.

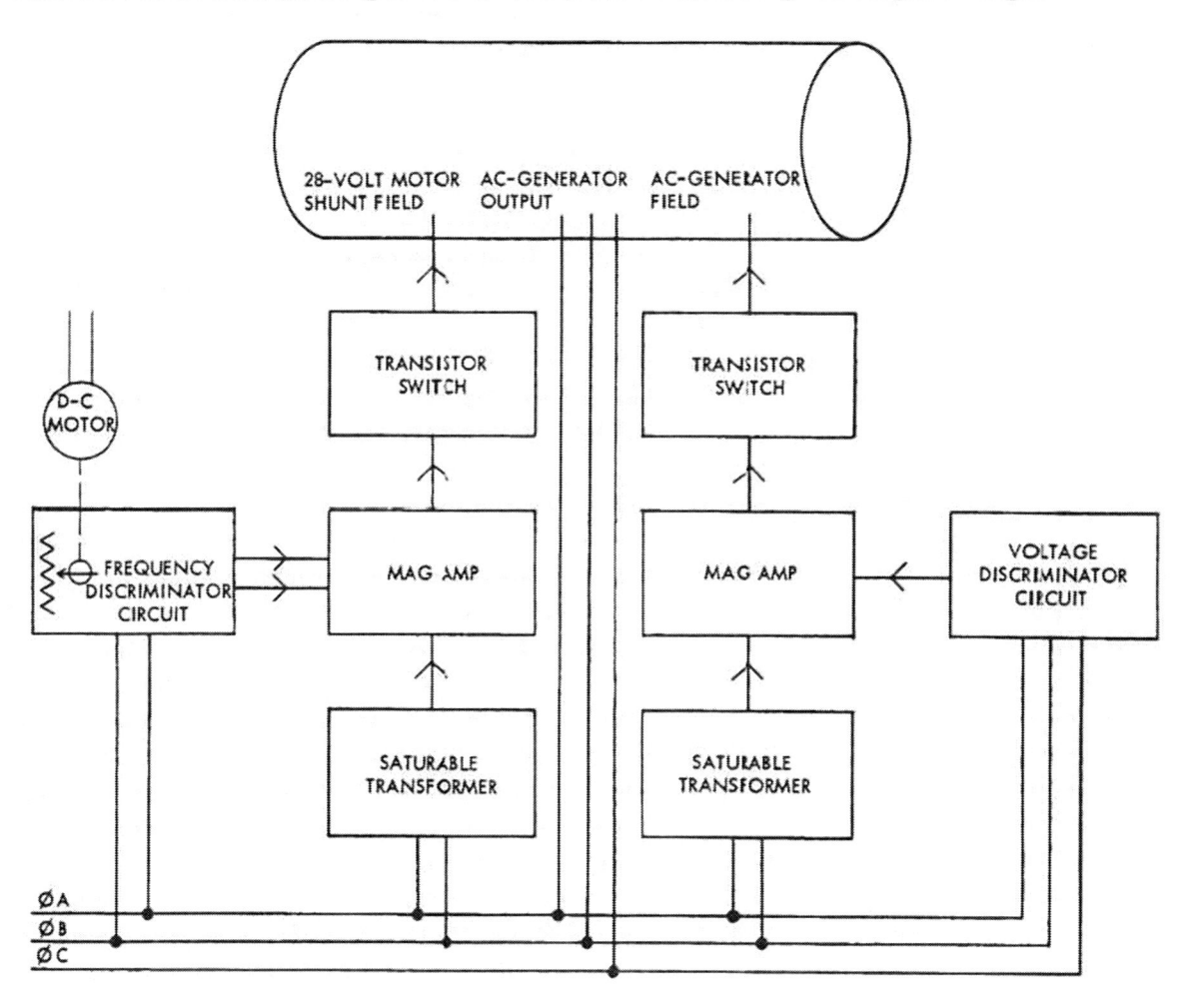

Figure IV-17 — Mod O Inverter Block Diagram

Phase II

The function of missile power components is to supply the a-c and d-c voltages necessary to sustain the missile in flight. In addition, the a-c supply furnishes voltage to the panels of the guided missile test station when the missile is being checked out.

The two 28-volt batteries, the 60-volt d-c power supply, and inverter 1 comprise the missile power supply system.

The two 28-volt batteries aboard the missile are remotely activated by the Guided Missile Test Station. Both batteries are enclosed in one common battery box and activated simultaneously. The "A" battery supplies power to the missile inverter; the "B" battery supplies power to the mechanical actuators and the relays of the missile system.

Both batteries contain 20 zinc-silver oxide cells with a total of 2000 cc potassium hydroxide as the electrolyte. Prior to battery activation, the electrolyte is stored in copper tubing that encircles the batteries. The activation signal from the ground fires two explosive squibs which build up pressure, forcing the electrolyte through the manifold into the battery cells. Activation time is approximately 5 minutes; the standby charge time is 12 hours. Battery "A" is connected to bus D9 and battery "B" is connected to bus +D8 at power transfer.

The 60-volt power supply provides 60 volts d.c. at a maximum of 2 amperes to the +F bus at power transfer. This voltage is used for the command potentiometers, beta feedback potentiometers, and the velocity and displacement potentiometers of the guidance computers.

The 115-volt, 60-cycle, 3-phase input is applied to the delta-connected power transformer. A full-wave rectifier is used in each phase, and the output is fed through a pi-type filter. A magnetic amplifier is utilized to provide voltage regulation.

Inverter 1 is identical to inverter 4, located in the power distribution trailer. In addition to the regulator circuits incorporated on the inverter, an inverter control panel is located in the guided missile test station. This panel, in conjunction with the precision generator, will compare the frequency of inverter 1 with the precision generator output voltage. If there is a difference of frequency, the inverter control panel will operate a switch which will apply a voltage to a frequency control motorized potentiometer in the inverter-regulator. This will cause the wiper of the potentiometer to move in one direction or the other, thus changing the resistance in the discriminator circuit. This will cause the frequency regulator circuit to change the current in the shunt winding of the motor, thus changing the frequency to that of the precision generator.

CHAPTER V
PROPULSION SYSTEM

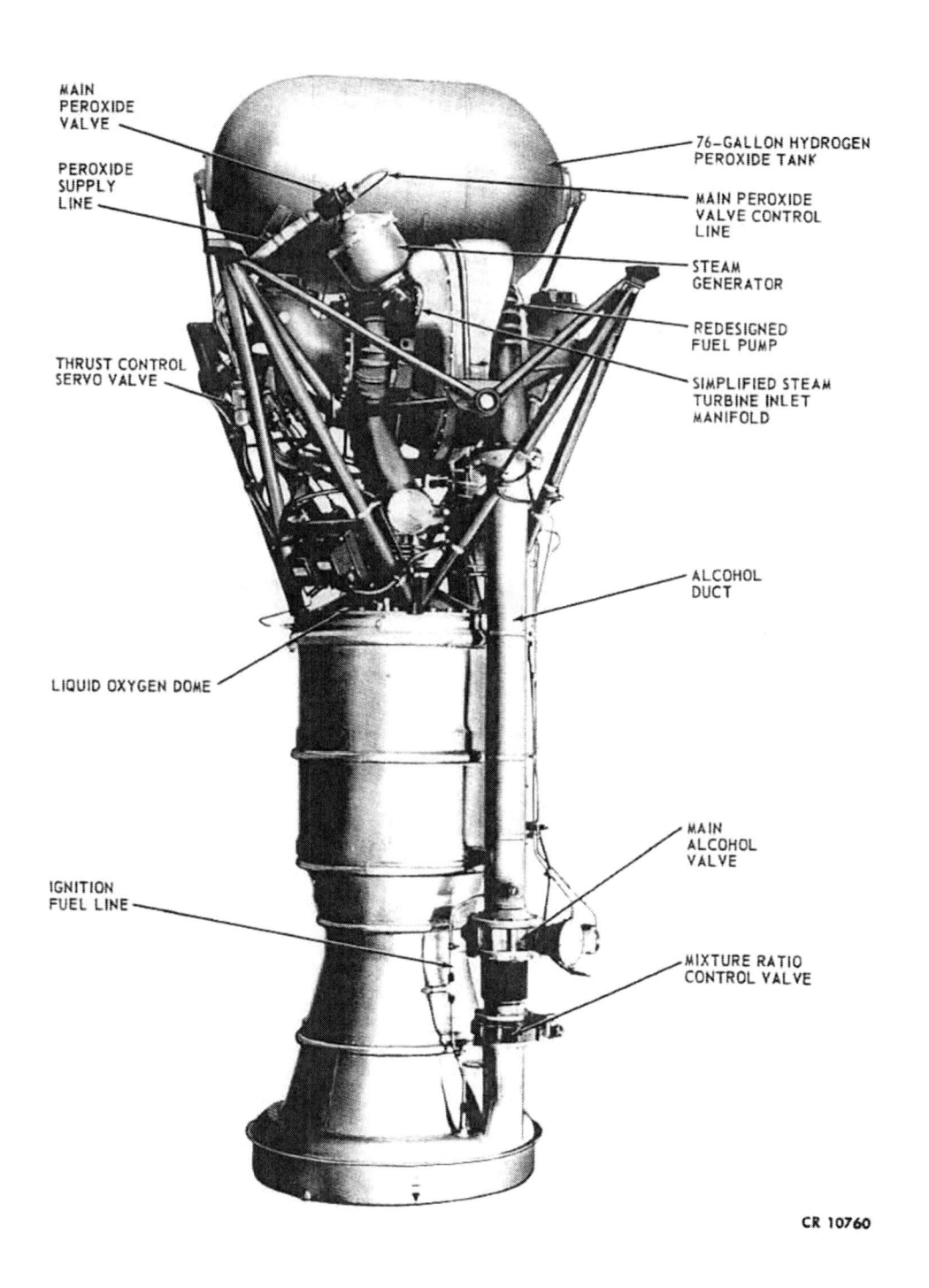

Figure V-1 — A-7 Rocket Engine

CHAPTER V
PROPULSION SYSTEM

GENERAL

The powerplant of the REDSTONE missile is a bipropellant liquid rocket engine manufactured by the Rocketdyne Division of North American Aviation Corporation.
This engine, which is an improvement over the German V2 engine, has a fixed thrust of about 78,000 pounds. This thrust can be generated for a period of 121 seconds, if necessary, but actual service use is less. The energy for this thrust is provided by a 75 per cent concentration of ethyl alcohol and LOX (the two propellants). These propellants are transferred from the missile tanks to the engine combustion chamber by high-pressure pumps which are driven by a steam turbine.
Although this engine is a fixed-thrust engine, small variations in thrust caused by ambient conditions can be corrected. This correction makes it possible to keep the thrust constant for guidance purposes and effectively removes a variable which would further complicate the system. The rocket engine, like other more familiar powerplants, needs several supporting systems for starting, stopping, and operating efficiently.

STEAM SYSTEM

This system is better known as the H_2O_2 system, or hydrogen peroxide system. Hydrogen peroxide, when decomposed rapidly, forms high-pressure, high-temperature steam. In the REDSTONE Missile the peroxide system is used to drive the propellant pumps by means of a steam turbine. The steam system has the following major components; a 76-gallon tank, a steam generator, a shutoff valve, a variable control valve, a pressurizing and venting valve, and an exhaust system.
When the launch start button is pushed, the pressurizing valve is opened and air is released at a pressure of 550 to 650 psi into the tank. The system is now static until the shutoff valve is opened. The peroxide flows through the variable control valve and into the steam generator, which is also known as the steam "pot". Inside the steam generator are potassium permanganate pellets. These pellets act as a catalyst, causing the peroxide to decompose rapidly into water and oxygen. This action releases a tremendous amount of heat and forms high-pressure, high-temperature steam. The steam is directed to the turbine, which in turn drives the fuel and LOX pumps. The pumps are attached to the same shaft. After passing through the turbine, the steam

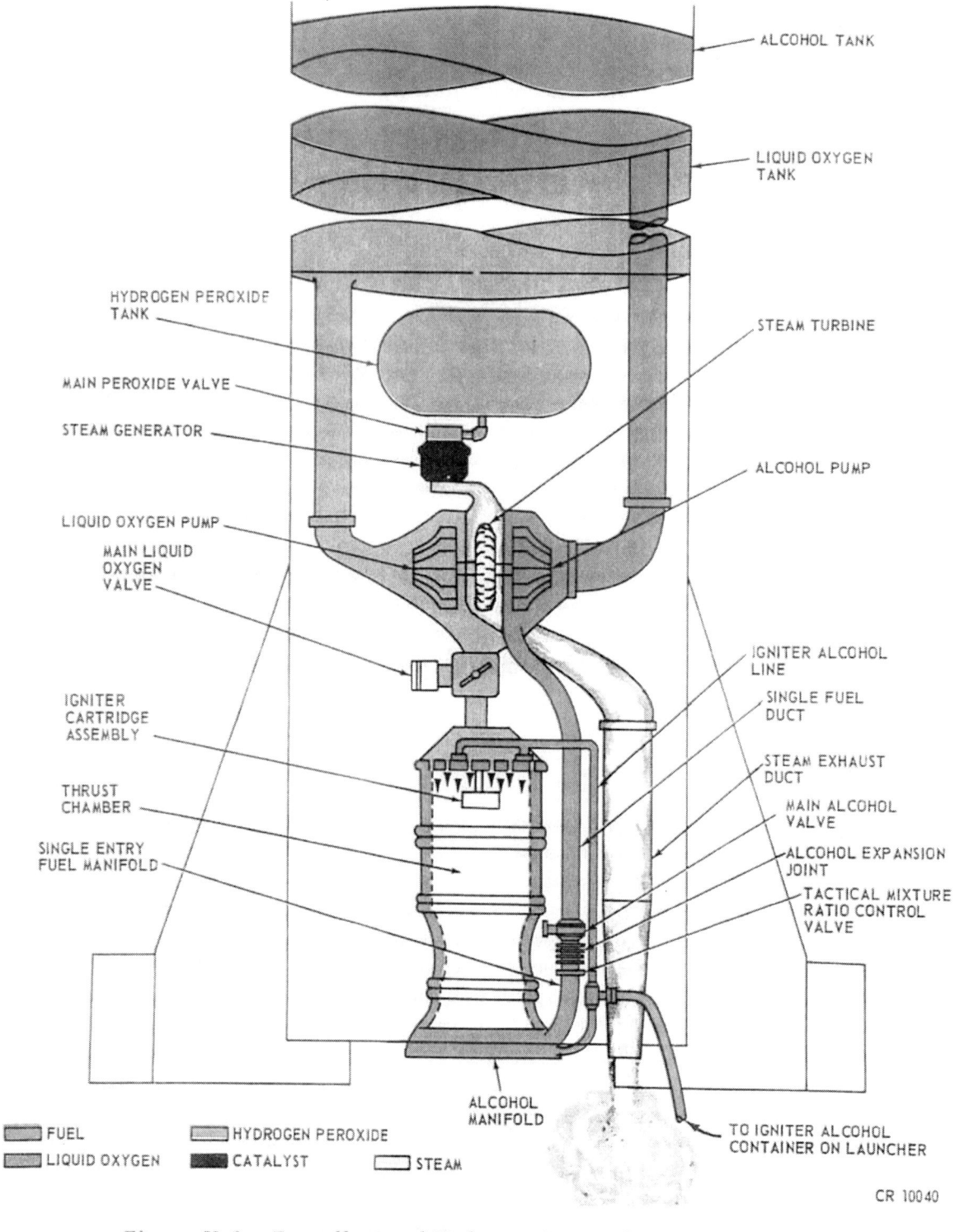

Figure V-2 — Propellant and Hydrogen Peroxide Flow Diagram

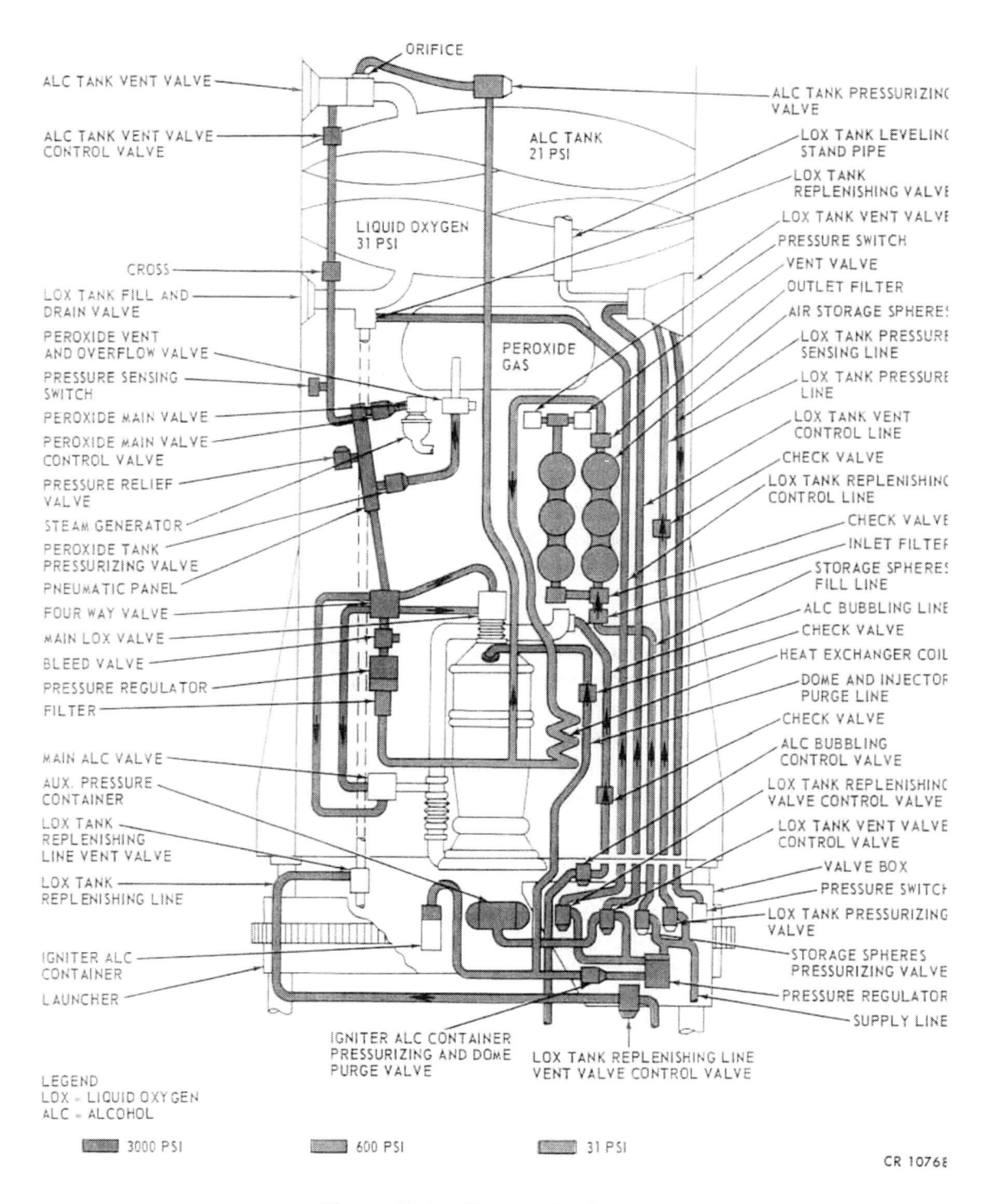

Figure V-3 – Pneumatic System

goes into the exhaust duct, where it vaporizes a small amount of LOX for the purpose of pressurizing the LOX tank of the missile, and to expand the main missile air supply as an economy measure. This is done through a unit called a heat exchanger, which is located in the exhaust duct. The steam is then sent overboard, adding a few hundred pounds of thrust to the missile.

PROPELLANT SYSTEM

The propellant system transfers the fuel and oxidizer from the missile tanks to the thrust chamber under pressure. The system consists of two tanks, two centrifugal-type pumps, two on-off control valves, a mixture ratio control valve, and the engine passages.

The two tanks are a part of the missile structure; the fuel tank is located immediately above the oxidizer tank. The fuel passes through the oxidizer tank (by way of a pipe) to the fuel pump. After leaving the fuel pump, the fuel is held from entering the engine by the closed main fuel valve. Following the main fuel valve is the mixture ratio valve. This valve is adjusted before flight to compensate for ambient effects on the fuel. Because alcohol density varies with temperature and pressure, varying amounts of fuel must be admitted into the engine in order to assure the proper thrust. From the mixture ratio valve, the fuel flows to a section, incorporated in the engine, known as the fuel manifold. This section collects the fuel and sends it up through the engine chamber walls. Fuel flowing through the walls of the engine serves two purposes. The primary purpose is to cool the engine chamber walls; no metal can withstand the 5,000°F temperature within the combustion chamber for very long. The fuel flows through the chamber very fast and absorbs so much heat that the temperature in the outside wall is maintained at about 125°F. The second purpose is to increase the thermal efficiency of the fuel by this heat absorption. Over-all system efficiency is thereby increased.

The engine passages lead to a part of the engine known as the injector. The injector is a steel plate that contains internal passages and rings. These rings have holes drilled at an angle so as to send a stream of fuel against a stream of fuel alternating with a ring with liquid oxygen doing the same. In this manner, the fuel and oxidizer is mixed at a given distance from the plate and forms a combustible mixture.

LOX flow is from the tank through the LOX pump and is stopped by the main oxidizer valve. From the valve, the LOX flows to the front (or top) of the engine and into the LOX dome and injector. The dome is just above the injector and serves as a manifold.

The injector performs two other functions. The outermost ring sends fuel down the thrust chamber inner wall and assists in the cooling of the thrust chamber. LOX and fuel alternate with the final ring (of twenty rings) discharging LOX. The other function is to assist in starting the engine. The center of the injector has a surface known as the ignition disc. Into this surface is fastened a pyrotechnic igniter which is similar to a flare. This surface also receives fuel from an external start tank. This fuel is directed through separate passages in the injector.

TURBOPUMP

The turbopump is the heart of the propulsion system. Without it, tank weight would be prohibitive because the tanks would have to be heavy in order to be sufficiently strong to permit adequate pressurization.

The turbopump consists of a turbine, a gear reduction section and two centrifugal pumps. These pumps are coupled to the turbine shaft and are driven at the same speed.

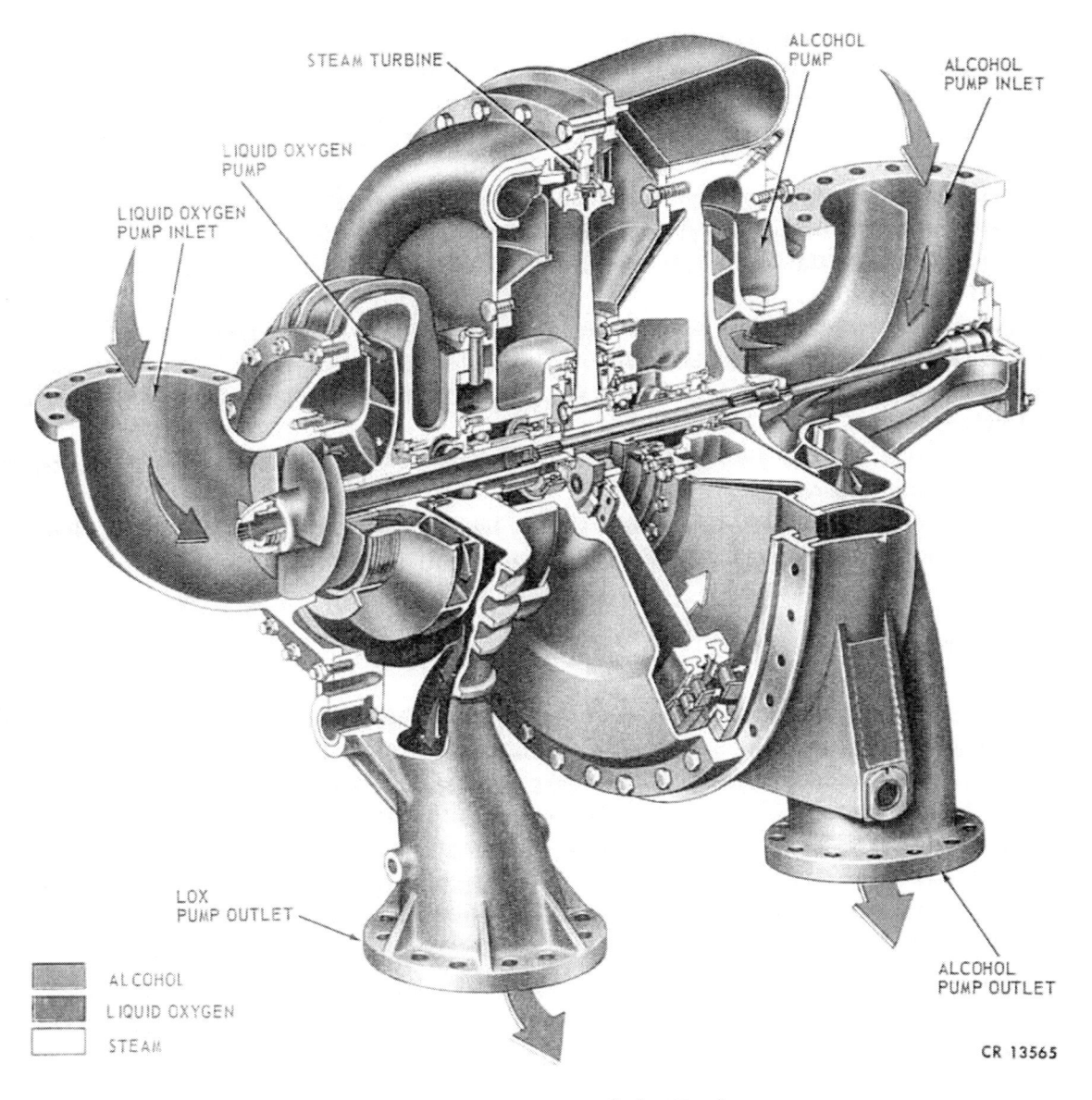

Figure V-4 — Operation of the Turbopump

The turbopump is the most critical item in the system. At the turbopump ambient temperatures of +700°F (steam) and -300°F (LOX) exist. Because of these extreme conditions, the turbopump must be well designed, even though it is in operation for only a short period of time.

PNEUMATIC SYSTEM

Engine starting and stopping must be controlled. The REDSTONE Missile uses a combination electro-pneumatic system to operate valves and to pressurize tanks.

An all-electrical system would require a bulkier, thus heavier, storage battery system and it would not assure reliable operation. An all-pneumatic system would require a large amount of tubing, which would make the system bulkier and more expensive.

A combination of the two has been found to be light, reliable, and inexpensive. Electrically operated solenoid valves are used to control the flow of high-pressure air to pressurize tanks and to operate the main fuel valve, oxidizer valve, and the peroxide shutoff valve.

The missile air supply is divided among three areas. The largest amount is stored in the section where the engine is located. This section supplies air to pressurize the fuel tank and the peroxide tank and to operate the main propellant valves and the shutoff valve.

The second largest amount is used to control the missile warhead and instrument section after separation.

The third supply is used to operate the air bearing system of the stabilized platform and to keep the instrument compartment at a constant pressure.

THRUST CONTROL

The thrust control system corrects for small thrust variations due to atmospheric conditions. This system makes use of the combustion chamber pressure to control thrust. (Thrust is a function of chamber pressure.) For example, if the chamber were designed to produce 75,000 pounds of thrust at 300 psig, the chamber would produce much less at 275 psig and much more at 325 psig. The only way the pressure can be changed in the REDSTONE system is by changing the quantity of propellants entering the chamber per unit time. Thus, if the chamber pressure is low, it is necessary to increase the flow rate of propellants into the chamber.

The thrust control system continuously monitors the chamber pressure and compares this pressure to a standard pressure preset into a thrust control amplifier. When the chamber pressure differs from the standard pressure, a signal is sent to the variable control valve in the steam system. This valve either increases or decreases the flow of peroxide to the steam generator which, in turn, increases or decreases steam flow.

As steam flow increases or decreases, the turbine speed also changes and, in turn, changes the speed of the turbopump. Changes in the speed of the turbopump cause the propellant flow rate to change which alters the chamber pressure and, hence, the thrust.

Low chamber pressure would cause a signal which would open the variable steam valve. This would increase the peroxide flow to the steam generator and increase the pump speed. More propellants per unit time would enter the chamber, bringing the chamber pressure up to the standard pressure set into the amplifier.

If the chamber pressure were too high, the system would decrease steam flow to slow the pumps and reduce the propellant flow rate. This would drop the chamber pressure, and, in turn, the thrust to the desired level.

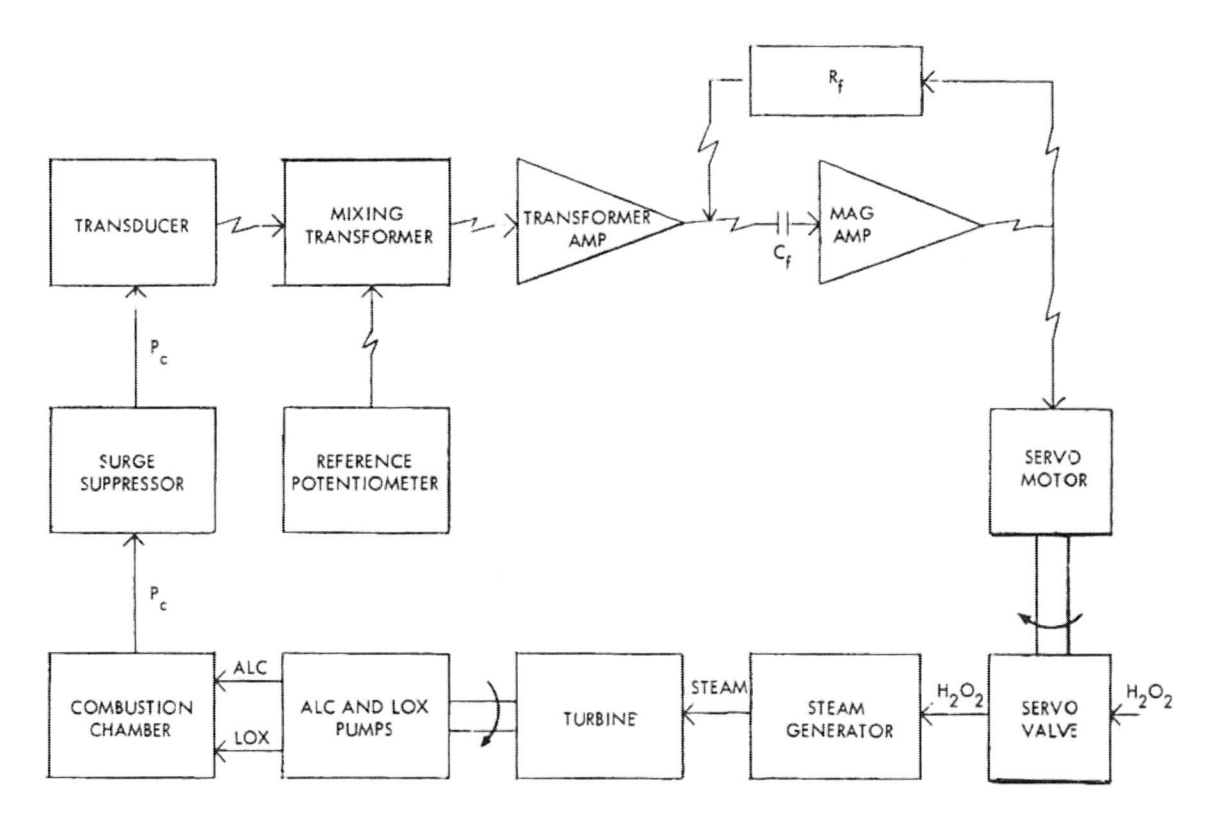

Figure V-5 – Thrust Controller

STARTING SYSTEM

The rocket engine has no moving parts and depends on an external system for starting. If the turbopump were started and the propellants were then ignited in the chamber, an explosion could result. Therefore, a small quantity of the propellants is sent to the chamber and ignited. The turbopump is then started, full thrust is obtained, and the missile is launched.

After all three tanks are pressurized, the pyrotechnic igniter in the injector is electrically fired. When this igniter fires, an electrical connection is broken, permitting the main oxidizer valve to open. This valve allows LOX to flow to the dome and through the injector and into the thrust chamber. When the oxidizer valve opens, it signals the solenoid that controls the starting fuel to admit fuel through the passages in the injector to the ignition disc. This fuel mixes with the LOX flowing into the chamber and oxygen rich ignition occurs.

When the fire in the chamber becomes hot enough, another wire located beneath the exhaust nozzle is burned through, which signals the peroxide shutoff valve and main fuel valve to open. The opening of the fuel valve is slowed down by a restriction placed in the line to permit the turbopump a small amount of buildup time. As soon as the turbopump reaches operating speed (about 0.3 seconds) the engine is in mainstage operation (between 90 and 100 per cent of rated thrust) and flight begins.

CUTOFF

When the missile has been in flight for a predetermined period of time, the guidance system sends a signal to the engine to shut down.

This is accomplished by first closing the peroxide shutoff valve and then closing the fuel and ozidizer valves.

The engine and tanks are no longer needed and are separated a short time later from the body unit. The thrust unit falls about ten miles short of the target.

CHAPTER VI
PROPELLANT SYSTEMS

This page has been left blank intentionally.

CHAPTER VI PROPELLANT SYSTEMS

GENERAL

A propellant may be defined as a solid, liquid, or gaseous material or various combinations of these materials whose heat of combustion is utilized to propel a missile.

In the selection of a propellant several factors, properties, and characteristics, including the following, must be considered:

Total impulse in lb-sec (I_T) = thrust in pounds (T) × duration in seconds (t). This is used in rating or comparing various propellants.

Specific impulse in lb-sec/lb (I_{sp}) = $\dfrac{\text{total impulse in lb-sec } (I_T)}{\text{weight of solid in lbs (W)}}$

This is used in rating or comparing solid propellants.

Specific thrust in lb/lb/sec (T_{sp}) = $\dfrac{\text{thrust in lb (T)}}{\text{weight rate of flow in lb per sec (W)}}$

This is a common method of comparing liquid propellants.

Density of a propellant is an important factor. A given weight of a dense propellant can be carried in smaller lighter tanks than the same weight of a lower density propellant. Liquid hydrogen, for example is high in energy and its combustion gases are light. However, it is a very bulky substance and requires large tanks. The dead weight of these tanks partially offsets the high specific impulse of the hydrogen propellant.

In engine operation, problems are sometimes created by chemicals that yield an excellent specific impulse. For example:

(a) Can the propellant be used adequately as a coolant for the hot thrust-chamber wall?

(b) Is the propellant sufficiently stable so that it can be safely stored and handled?

Most propellants are corrosive, flammable, and/or toxic. A propellant that gives good performance is usually a highly active chemical. Many propellants are highly toxic, to a greater degree even than most war gases. Some propellants are so corrosive that only a few special materials can be used to contain them. Some propellants burn spontaneously upon contact with air or upon contact with an organic substance, or in certain cases, upon contacting most common metals.

Since some propellants are used in very large quantities, the availability of raw materials must be considered. Also, in some cases, an entire new chemical plant must be built in order to obtain adequate amounts of a propellant.

In the chemical-type rocket engine two general types of propellants are used; the liquid propellant and the solid propellant.

Two general types of the solid propellant are in use: the double-base propellant and the composite type. The double-base propellant consists of nitrocellulose and nitroglycerine, plus additives in small quantities. There is no separate fuel and oxidizer. Extrusion methods are usually employed in the manufacture of this type of solid propellant. However, casting has been employed.

The composite type of solid propellant uses separate fuel and oxidizer which are intimately mixed in the solid grain. The oxidizer is usually ammonium nitrate, potassium chlorate, or ammonium chlorate and often comprises as much as four-fifths or more of the whole propellant mix. The fuels used are hydrocarbons, such as asphaltic-type compounds, or plastics.

Solid propellants offer the advantage of minimum maintenance and instant readiness. However, the solids usually require carefully controlled storage conditions and offer handling problems, particularly in the very large sizes. Protection from mechanical shocks or abrupt temperature changes that may crack the grain is essential.

Liquid propellants are divided into three general classes: Fuels, oxidizers, and monopropellants. Theoretically, the most effective fuel component is hydrogen, and the most effective oxidizing agents are fluorine or ozone.

Liquid ammonia, the alcohols, aniline mixtures, hydrazines, hydrogen, and various petroleum products and derivatives are used or are theoretically possible to use as propellants.

In the class of oxidizers, the following are used or considered for use: LOX, fuming nitric acids, fluorine, chlorine trifluoride, ozone, ozone-oxygen mixtures, and concentrated hydrogen peroxide.

Ethylene oxide, hydrazine, hydrogen peroxide, and nitromethane are monopropellants. It follows that hydrazine and hydrogen peroxide may be utilized as a fuel and oxidizer, respectively, as well as a monopropellant.

Certain propellant combinations are hypergolic; that is, they ignite spontaneously upon contact of the fuel and oxidizer. Others require an igniter to start burning, although they will continue to burn when injected into the flame of the combustion chamber.

In general, the liquid propellants in common use yield specific impulses superior to those of available solids. On the other hand, they require more complex propulsion systems to transfer the liquid propellants to the combustion chamber.

Certain unstable liquid chemicals, which under proper conditions will decompose and release energy, have been tried as rocket propellants. Their performance, however, is inferior to that of bipropellants and they are of interest only in specialized applications.

Outstanding examples of this type of propellant are hydrogen peroxide and ethylene oxide. Occasionally, a separate propellant is used to operate the gas generator which supplies the gas to drive the turbopumps of liquid rockets.

Table VI-I – SPECIFIC IMPULSE OF SOME TYPICAL CHEMICAL PROPELLANTS

Propellant combinations:	Isp range (sec)
Monopropellants (liquid):	
Low-energy monopropellants	160 to 190.
Hydrazine	
Ethylene oxide	
Hydrogen peroxide	
High-energy monopropellants:	
Nitromethane	190 to 230.
Bipropellants (liquid):	
Low-energy bipropellants	200 to 230.
Perchloryl fluoride – Available fuel	
Analine – Acid	
JP-4 – Acid	
Hydrogen peroxide – JP-4	
Medium-energy bipropellants	230 to 260.
Hydrazine – Acid	
Ammonia – Nitrogen tetroxide	
High-energy bipropellants	250 to 270.
Liquid oxygen – JP-4	
Liquid oxygen – Alcohol	
Hydrazine – Chlorine trifluoride	
Very high-energy bipropellants	270 to 330.
Liquid oxygen and fluorine – JP-4	
Liquid oxygen and ozone – JP-4	
Liquid oxygen – Hydrazine	
Super high-energy bipropellants	300 to 385.
Fluorine – Hydrogen	
Fluorine – Ammonia	
Ozone – Hydrogen	
Fluorine – Diborane	
Oxidizer-binder combinations (solid):	
Potassium perchlorate:	
Thiokol or asphalt	170 to 210.
Ammonium perchlorate:	
Thiokol	170 to 210.
Rubber	170 to 210.
Polyurethane	210 to 250.
Nitropolymer	210 to 250.

Table VI-I – SPECIFIC IMPULSE OF SOME TYPICAL CHEMICAL PROPELLANTS (Continued)

Propellant combinations:	Isp range (sec)
Oxidizer-binder combinations (solid): (continued)	
Ammonium nitrate:	
Polyester	170 to 210.
Rubber	170 to 210.
Nitropolymer	210 to 250.
Double base	170 to 250.
Boron metal components and oxidant	200 to 250.
Lithium metal components and oxidant	200 to 250.
Aluminum metal components and oxidant	200 to 250.
Magnesium metal components and oxidant	200 to 250.
Perfluoro-type propellants	250 and above.

OXIDIZER SYSTEM

Characteristics and Properties of Oxygen

Oxygen may exist as a solid, liquid, or gas. These states are determined by the temperatures and pressures under which it is handled. Oxygen may be liquefied if cooled below a temperature of -297.35°F at atmospheric pressure. By increasing the pressure, oxygen may exist as a liquid at temperatures above -297°F. The critical temperature of oxygen is -182°F, it will not remain a liquid above this temperature regardless of the pressure applied.

Symbol – O, formula – O_2

Atomic weight – 16

Valence – 2

Concentration in air – 20.99% sea level

Color of LOX – pale blue

Viscosity (LOX) – nonviscous (water-like fluid)

Normal freezing point – (-219°C) (-362°F) sea level

Normal boiling point – (183°C) (-297°F) sea level

Critical temperature – (-119°C) (-182°F) sea level

Critical pressure — 47.7 atmospheres (730.6 psi)

Density:

   Gas   — 0.08305 lb/cu ft
   Liquid — 71.5 lb/cu ft, 9.69 lb/gal

Gaseous oxygen is approximately 1.103 times heavier than air.

LOX is approximately 1.14 times heavier than water.

Gas constant R = 48.31 ft-lb/lb °R

Specific Heat — 0.4 Btu/lb/°F

Heat of Vaporization — 92 Btu/lb

Surface Tension — 18.3 dynes/cm

LOX is insoluble with all common solvents, since it freezes them on contact.

In its gaseous state it is a strong oxidizing agent.

The extreme cold temperature of LOX tends to make metals, plastics, rubber, and most other materials very brittle.

Oxygen is nontoxic.

Specification — Type II, Grade "A" of MIL-O-8069

LOX is an explosive hazard if subject to shock or ignition when contaminated with organic materials such as oil, grease, carbon black, paper, wood, cork, gasoline, JP fuels, kerosene, and metal in the form of powder or shavings.

In the absence of organic contamination LOX is considered stable and its vapors create no particular hazard. However, it supports combustion when in the vicinity of, or enclosed with, combustible material.

LOX is always -182°F or lower in temperature (depending on pressure), is pale blue in color, and flows like water. Under no condition should it be restricted in a given space. For example, one cubic foot of LOX represents over 800 cubic feet of gaseous oxygen at atmospheric pressure and would build up to approximately 112,000 psi if confined to the original cubic foot of space.

LOX is dangerous, and there are a few rules that must be followed to assure against accidents. Possible danger is based on three general characteristics of LOX:

The rate of combustion of most materials can be greatly increased by the presence of pure oxygen.

Human contact with LOX or uninsulated lines at a temperature of -297°F can result in severe frostbite. Some types of material and equipment vulnerable to freezing conditions can be damaged easily.

LOX, if confined, will eventually evaporate and build up a tremendous pressure which will result in the rupture of the tank in which it is stored.

The ground equipment used in the oxidizer system consists of LOX generating and production equipment, a storage container, transportation vehicles, and transfer equipment.

The Engineering Company is equipped with the LOX generating facilities (an air supply semitrailer and an air separation semitrailer). The air supply semitrailer contains the compressed air supply assembly powerplant which is powered by four diesel engines, each of which drives four 4-stage air compressors.

The air separation semitrailer contains the oxygen-nitrogen separation assembly, which consists of the heat exchangers, air dryers, refrigeration system distillation column, carbon dioxide filters, and an electric generator. Both semitrailers are towed by standard M-52 truck tractors.

LOX from the generating plant is stored in tanks that have a capacity of 70,000 pounds each. The storage tank is basically a large thermos bottle; that is, it has double wall (inner and outer tank) with a vacuum-insulated space between. It is skid mounted and complete with pump, valves, switches, indicators, and controls for transfer of LOX. The storage tanks are equipped with safety valves and rupture disks for protection against excessive internal pressure.

LOX from the storage tanks is transported and stored at the launch site in two 9-ton, tank-type semitrailers towed by standard M-52 truck tractors. The transporters are similar in construction to the storage tank. The offset design of two different diameters gives optimum weight distribution and provides a low center of gravity with a fifth wheel. A compartment at the rear of the trailer houses the transfer equipment necessary to transfer LOX to the missile tank at rates up to 150 gpm against head pressures up to 75 psi.

The main LOX components of the REDSTONE missile system are: the LOX fill and drain valve, the LOX container or tank, the LOX pump, the main LOX valve, the LOX dome, the injector and the combustion chamber. Since LOX evaporates (boils off), it is necessary to replace this loss by means of a replenishing valve. Also, excessive pressure buildup is vented to the atmosphere by means of a vent valve. Miscellaneous equipment includes such items as a transfer hose, a "Y" connector, a fueling ladder, couplings, and other materiel.

Because of size, weight, and structural considerations, propellants are transferred to the missile only when the missile is in the vertical position.

Filling Procedure

LOX is delivered to the firing site by the two 9-ton semitrailers and positioned near the missile. The trailers are connected to the missile fill and drain port by means of hoses, the "Y" connector, the fueling ladder, and other equipment. After the hookup is completed, precooling of the lines and pumps is started by a gradual flow of LOX through the delivery system. When precooling has been accomplished, the pumps are activated and the transfer of LOX is accomplished. One semitrailer starts the pumping operation first, and after an interval of three or four minutes, the second semitrailer commences to pump. This procedure assures that a sufficient amount of LOX will be reserved in the second semitrailer for replenishing.

When the filling operation (filling weight is 25,430 pounds) has been completed, hoses and accessories are disconnected and secured. One semitrailer is withdrawn from the site while the trailer with the reserve supply of LOX moves to the replenishing position (150 feet from the missile). A replenishing line is installed, and the LOX is topped into the missile tank at regular intervals to replenish the loss from evaporation until the missile fire command is given.

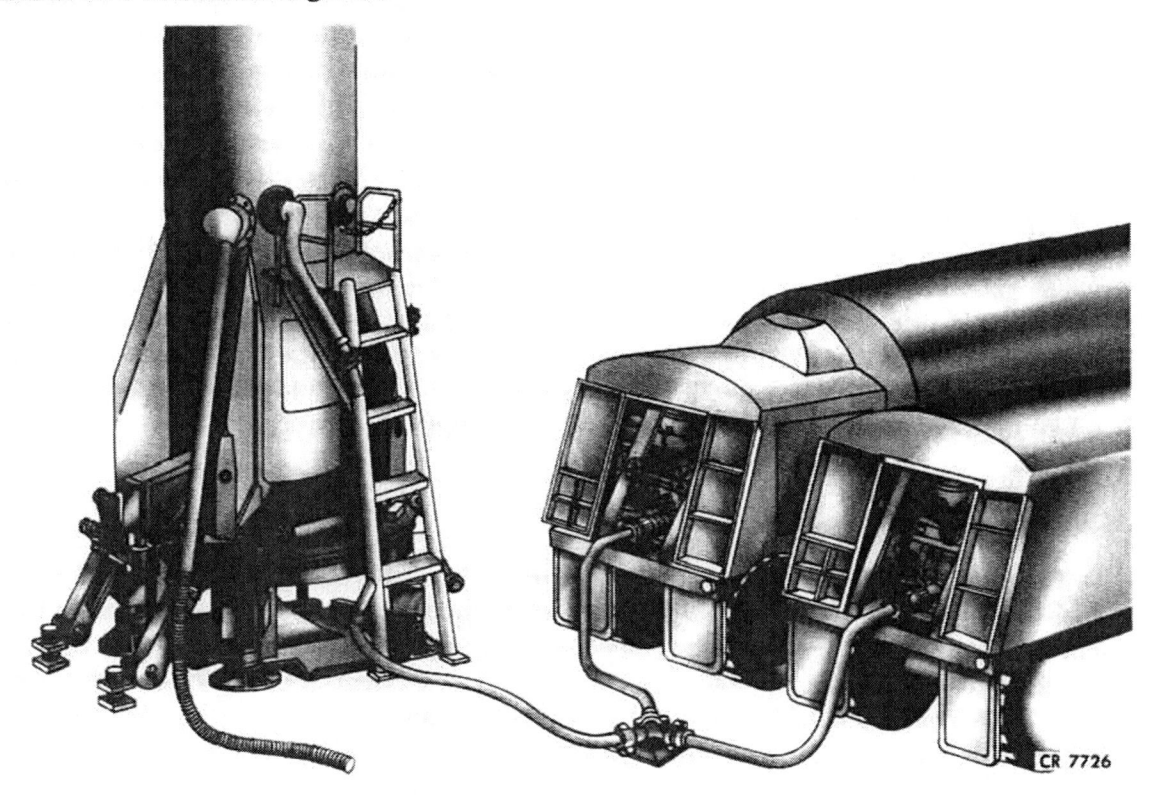

Figure VI-1 — LOX Filling

FUEL SYSTEM

The REDSTONE Missile uses a denatured ethyl alcohol and water mixture (75 per cent alcohol, 25 per cent water) as a fuel. This fuel is mixed with LOX (oxidizer) in the thrust chamber and burned to produce thrust for propelling the missile.

Ethyl alcohol (C_2H_5OH), or grain alcohol, is a colorless, chemically stable, flammable liquid. It has a characteristic odor and is not sensitive to shock. Ethyl alcohol is relatively inert and does not react to any great extent with the metal of valves, piping, or tanks. Other properties of ethyl alcohol are:

Boiling point (°C)	78.4
Flash point (°C, enclosed)	12
Auto-ignition temperature (°C)	426
Specific gravity at 20/4°C	0.789
Vapor density	1.59
Vapor pressure at 25°C (mm Hg)	50

The alcohol is shipped and stored in 55-gallon (expendable) steel drums. The alcohol is transferred from the drums to the alcohol semitrailer, the basic function of which is to transport alcohol and inert fluid to the firing site and to pump alcohol and inert fluid into the REDSTONE Missile fuel system. The 3,000-gallon, 2-wheel, XM388 alcohol semitrailer consists of a 3,000-gallon aluminum tank, and a pumping compartment which contains a 20-gallon stainless-steel tank (for inert lead fluid) and two pumping units. The semitrailer is mounted on a single-axle chassis.

The alcohol tank is an elliptically-shaped, one-compartment tank with a capacity of 3,000 gallons plus 3 per cent expansion. A 1-inch-thick rubber insulation covers the entire tank, with the exception of the pumping compartment. The insulation helps to keep the alcohol at the proper temperature. The tank contains the necessary vents, fusible plugs, a manhole and fill cover assembly, a capacity indicator, a strainer, and other items necessary for the safety and operation of the tank.

Fuel Transfer System

The primary function of the fuel transfer system is to deliver strained and metered fuel to the missile. The system is designed to perform a variety of associated fueling operations including: evacuating the fueling hose, filling the semitrailer tank from alcohol drums, mixing or recirculating the tank contents, and servicing the missile with fuel from an outside source. It is also capable of pumping into the semitrailer from an outside source through the gravity drain valve, pumping out of the semitrailer through the meter bypass, draining into or out of the semitrailer gravity drain valve, draining by gravity flow from the missile back to the semitrailer tank, and pumping from the missile into the semitrailer through the eductor system. The transfer equipment consists of a centrifugal pump coupled to an explosion-proof motor, a fuel metering chamber and register, gages and instruments, and other valves and associated piping used to direct the fuel flow.

The fuel transfer pump is mounted on the end of the motor shaft and is designed to deliver 250 gpm at a static head of 43 feet. The pump is a centrifugal, non-self-priming type and is equipped with a mechanical seal. The motor is rated at 15 hp at 3000 rpm and operates on 208-volt, 60-cps, 3-phase power.

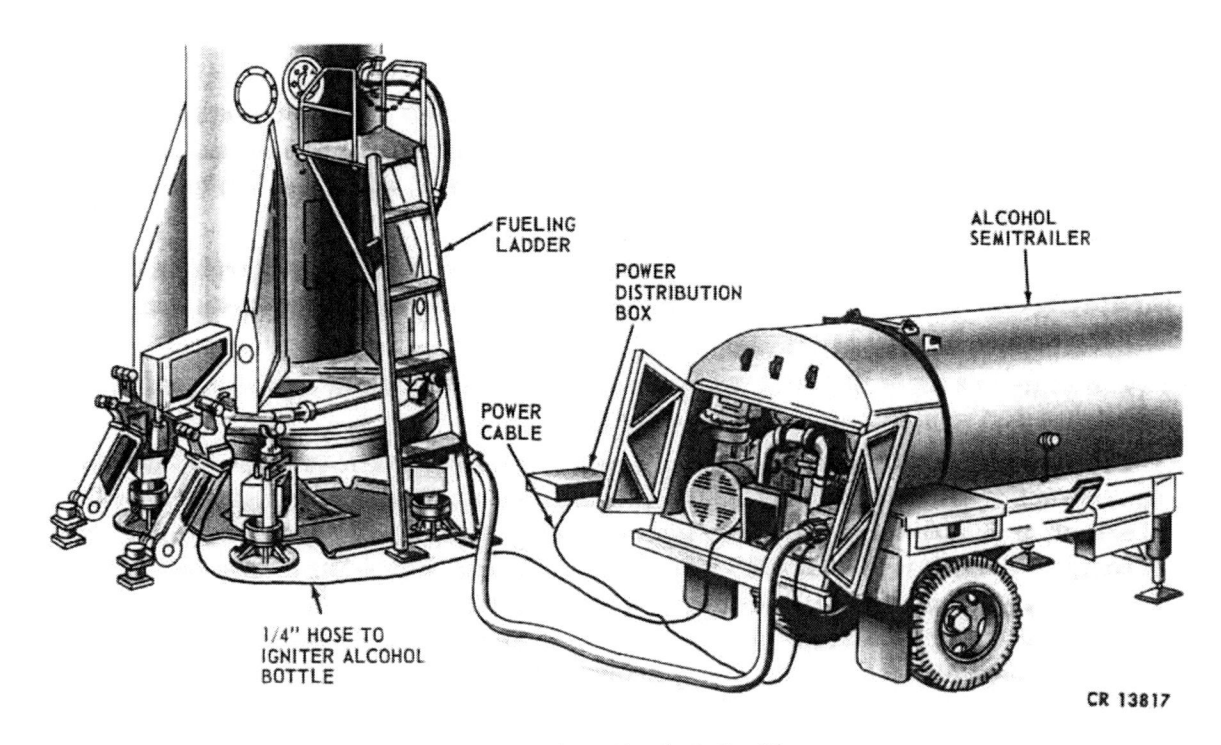

Figure VI-2 — Alcohol Fueling

The metering system consists of the meter register, the auto-stop valve, the metering chamber, and the air release and strainer. The purpose of the system is to accurately record the amount of fuel being pumped into the missile.

Other equipment such as gages, instruments, valves, and associated piping are used to control and direct the fuel flow but these items are considered to be standard equipment.

Tank Heater System

The tank heater system is designed to heat the alcohol-water fuel solution prior to fueling the missile. Initial heating takes place in the support area. Any additional heating required to maintain the prescribed fuel temperature would take place at the firing site while the alcohol semitrailer is in standby position. In some cases, the entire heating operation may be performed at the firing site. Circumstances will dictate the procedure used, and the final choice is the prerogative of the firing command.

The heater system is electrically powered by a 208-volt, 3-phase, 60-cps supply received from an external power source. The system consists of twelve 4.5-kilowatt immersion-type electrical heating elements, six on each side of tank; two magnetic starters, located in the heater distribution box; four thermostats, two on each side of the tank; two safety systems; and an indicator light unit located in the rear of the pump compartment.

Inert Fluid Transfer System

The function of the inert fluid system is to deliver inert fluid to the missile.

The inert fluid transfer equipment consists of a centrifugal pump and explosion-proof motor mounted on a common shaft, a stainless-steel tank, strainers and valves, a hose reel, and other plumbing necessary to direct the flow.

The inert fluid tank is a stainless-steel, rectangular, single-compartment tank with a capacity of 20 gallons. It is suspended from the top of the pump compartment by three aluminum straps.

The seal-less inert fluid pump is on a common shaft with the explosion-proof motor and delivers 5 gpm at a minimum 40-foot head.

The semitrailer arrives at the firing site along with the various other ground support equipment. The alcohol semitrailer remains in standby status until after vertical checkout of the missile. Preparations are then made for fueling the missile. These preparations consist of: installing the missile fueling ladder, installing fueling valves and hoses, emplacing the air heater, and providing a lithium chloride lead start which consists of pumping lithium chloride into that portion of the missile alcohol system below the main alcohol valve in order to provide smoother ignition when the missile is fired. Fueling is then begun. After fueling the missile, the alcohol semitrailer is moved out of the firing area.

The main items of the missile fuel system are: the alcohol fill and drain valve, the alcohol tank, the pump, main alcohol valves, the injector, the combustion chamber, and other associated components such as the vent valve, the pressurizing valve, check valves, piping, and couplings.

The alcohol tank, which makes up the forward portion of the missile center section, is filled and drained through the mechanically actuated alcohol tank fill and drain valve. This valve is mounted on the missile skin near the aft bulkhead of the center section and is connected to the main alcohol supply ducting by an expansion joint. The valve is composed of a cast aluminum-alloy body, an O-ring type seat which is spring-loaded to the normally closed position, a seat ring, and an adapter flange.

The alcohol pump forces the flow of alcohol from the alcohol tank to the rocket engine combustion chamber at a delivery pressure of 420 psi. The fuel pump and turbine are coupled to, and driven by, a common shaft. The alcohol pump is of the single-entry, centrifugal type, utilizing radial flow impellers. The impellers incorporate balance ribs on the inboard side to facilitate hydraulic balancing of end thrust by the adjustment of rib-to-housing clearances. The alcohol pump has a nominal delivery of 1333 gpm.

The main alcohol valve controls the flow of fuel to the rocket engine combustion chamber. It is actuated pneumatically by a four-way solenoid valve. The valve is installed between the thrust chamber alcohol manifold and the alcohol ducts.

The injector disperses alcohol into the rocket engine combustion chamber in small streams and mixes with like streams of LOX also introduced by the injector.

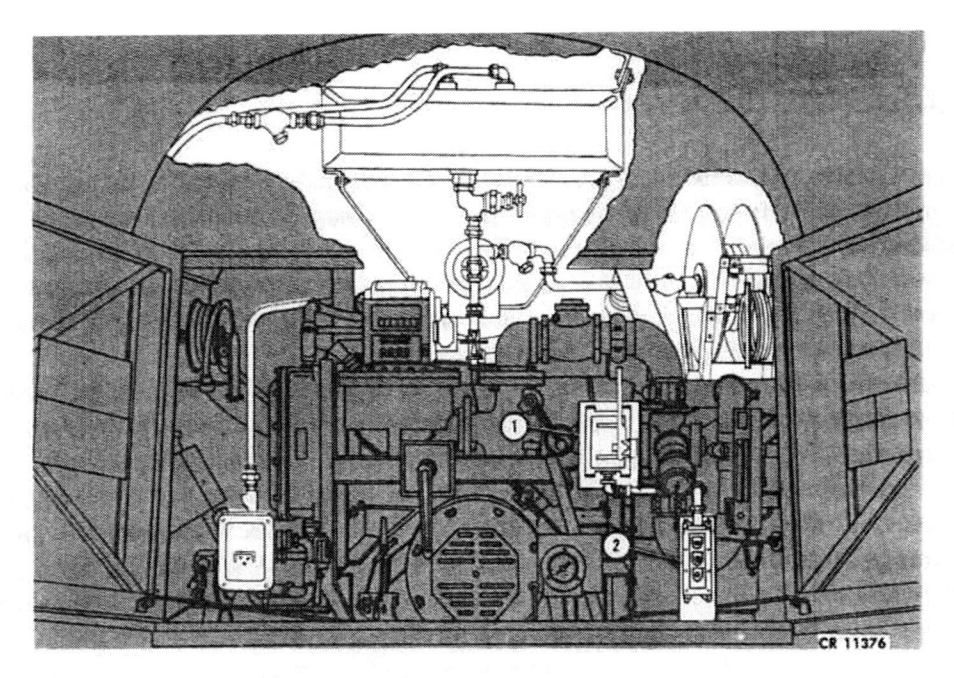

Figure VI-3 — Inert Fluid System

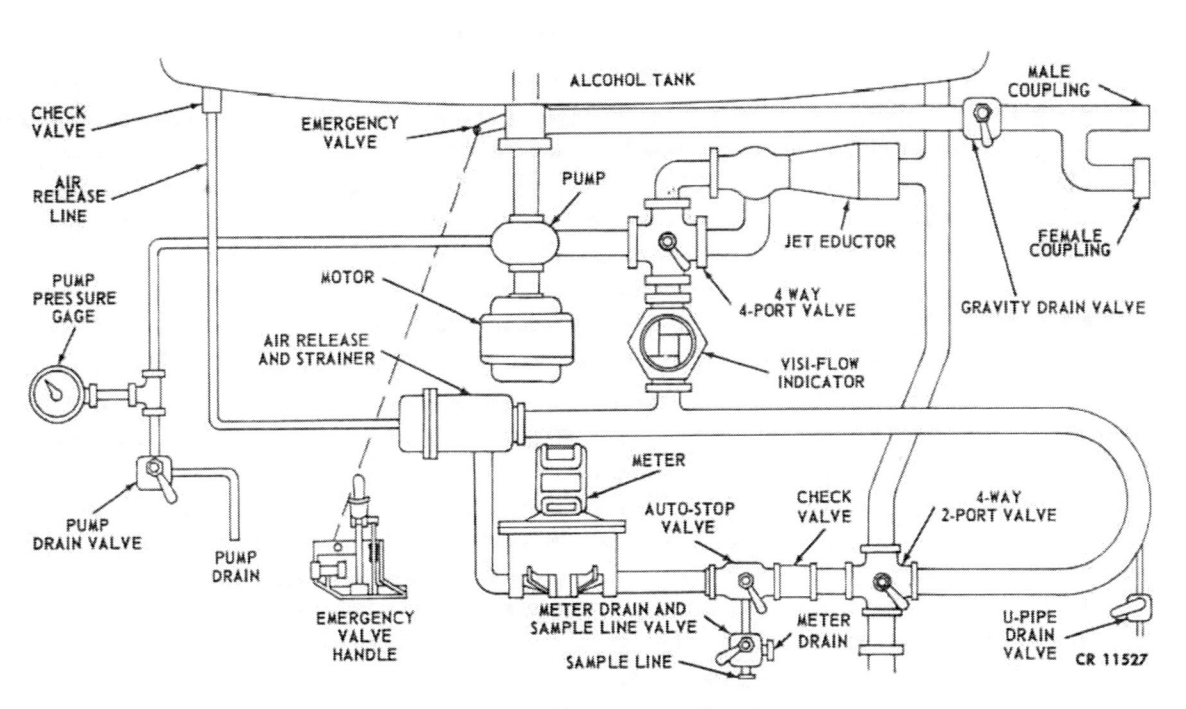

Figure VI-4 — Fuel Transfer System

VI-11

HYDROGEN PEROXIDE SYSTEM

The hydrogen peroxide system provides the power for operating the turbopump assembly that delivers propellants to the rocket engine combustion chamber.

Hydrogen peroxide is a colorless liquid. In the pure state, it is extremely stable; however, any impurities introduced during manufacture, shipment, storage, or handling will reduce this stability. Metals such as iron, copper, chromium, and their salts, will decompose hydrogen peroxide on contact.

Hydrogen peroxide itself is nonflammable. However, it is a strong oxidizer and, if allowed to remain in contact with readily oxidizable organic materials, may cause spontaneous combustion. In addition, hydrogen peroxide solutions are catalytically decomposed by many metals and their salts, ordinary dirt, ferments, enzymes, and other substances. This decomposition liberates oxygen, which will promote the combustion of flammable materials. Concentrated solutions are powerful oxidizing agents that can rust iron and steel. These solutions can furnish oxygen and heat to burn combustibles such as wood, leather, paper, cotton, and wool.

All materials which must come into contact with concentrated hydrogen peroxide are chosen very carefully. Storage tanks and pipe lines should be made of 99.6 per cent or 2S aluminum. Pipe should be bent and flange welded to avoid objectionable screw-on fittings. Gate valves are recommended and should contain 99.6 per cent aluminum, 2S, or 43S alloy. Pumps should be made of 43S aluminum or type 316 stainless steel. All tanks and containers must be treated by a special pickling procedure before use. During all operations, the prescribed protective clothing is worn.

Hydrogen peroxide is received and stored in special 86-gallon drums, which are transported to the firing site by the hydrogen peroxide truck. The hydrogen peroxide truck also supplies hydrogen peroxide to the missile by means of a pump, hoses, and other equipment. The truck is a modified 3/4-ton M-37 cargo truck. The modifications to the truck include a monorail system, an electrically operated pumping unit, an overflow tank, and the necessary fittings, hoses and cables.

The monorail assembly consists of an I-beam welded to, and supported by, two A-frames which are bolted to the sides of the truck. The purpose of the monorail assembly is to provide a support for the chainfall assembly when it is used to handle the drum. The monorail extends beyond the end of the truck to allow the hydrogen peroxide drum to be lifted from the ground. The chainfall assembly is used to raise and lower the drum and to traverse the length of the monorail.

There are three cover assemblies used with the hydrogen peroxide drum: the lower collection ring cover assembly, the upper collection ring cover assembly, and the drum cover assembly. The cover assemblies provide insulation and air space around the drum to aid in maintaining the hydrogen peroxide at the prescribed temperature (65°F to 85°F). The air space allows for the circulation of hot air (for heating) or gaseous nitrogen (for cooling). A heater assembly, in conjunction with the cooling system, regulates the temperature of the hydrogen peroxide.

The hydrogen peroxide pump is a rotary, positive-displacement type, capable of delivering 8 gpm; the pump serves two functions in the missile system: 1) to pump hydrogen peroxide from the drum into the missile, and 2) to recirculate hydrogen peroxide through the drum for even temperature distribution.

An overflow tank assembly (6-gallon capacity) is used to catch the overflow of hydrogen peroxide from the missile during missile loading operations.

After servicing, all equipment except the overflow tank and hose is returned to the hydrogen peroxide truck, which then moves to a safe distance from the firing site. The overflow tank and hose are moved from the firing site at a specified time just prior to missile firing.

The main components of the missile peroxide system are: the fill and drain valve, the tank, the thrust control servo valve, the main peroxide valve, the vent and overflow valve, the control valve, and the steam generator.

The hydrogen peroxide tank, which is mounted to the thrust frame at the forward end of the rocket engine, is filled and drained through the mechanically actuated peroxide fill and drain valve. The tank has a capacity of 75 gallons.

The main peroxide valve is a pneumatically operated, poppet-type valve that is spring loaded to a normally closed position.

The steam generator is a steel body and cover assembly which is copper-coated and nickel-plated to resist corrosion. The generator is mounted on the steam inlet manifold of the turbopump and contains a pellet-type decomposition catalyst (potassium permanganate). Hydrogen peroxide, which enters the top of the generator under supply tank pressure, is sprayed by the injector through the diffuser plate and into the catalyst, where it decomposes into steam. The steam is then directed from the bottom of the generator into the turbopump manifold.

The electrically operated, gate-type, thrust control servo valve, installed in the peroxide supply line upstream of the main peroxide valve, controls motor chamber pressure by automatically throttling the peroxide flow to the steam generator. The reduction in peroxide flow regulates the pumping rate of the turbopump.

After passing through the steam turbine, the steam enters an exhaust duct which directs it through the heat exchanger. The steam heats the coils in the heat exchanger and then passes overboard through the exhaust duct.

The pumping system recirculates hydrogen peroxide prior to servicing the missile, and pumps hydrogen peroxide into the missile.

All the hydrogen peroxide in the drum must be at the same temperature before it is pumped into the missile. The purpose of recirculation is to mix the hydrogen peroxide and thereby maintain a consistent temperature in the drum.

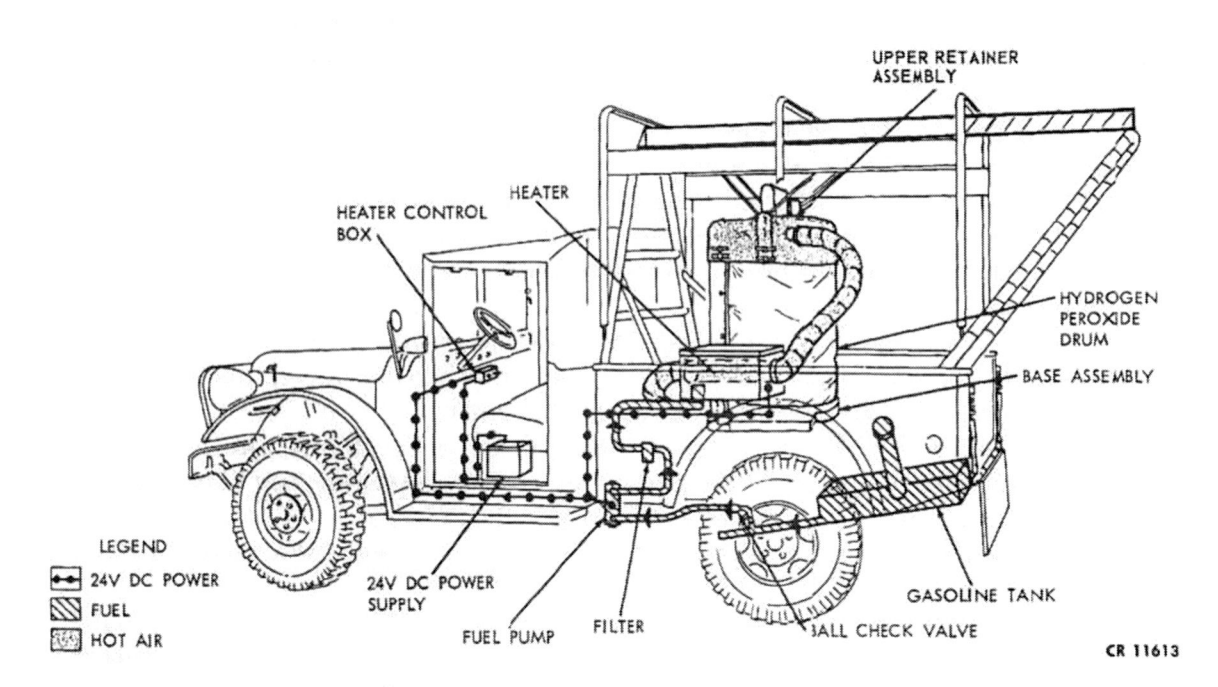

Figure VI-5 — Drum Heating System

OUTPUT
  ON-HI . . . . 30,000 BTU/hr
  ON-LO . . . . 12,000 BTU/hr

VOLTAGE REQUIRED
  Maximum. . . . . . . 28 vdc
  Minimum. . . . . . . 18 vdc

FUEL   any grade gasoline

CURRENT CONSUMPTION
  Starting . . . . 14 amperes
  Operating. . . . . . . 7 amperes

FUEL CONSUMPTION
  Maximum. . . . . . . 0.44 gph
  Minimum. . . . . . . 0.20 gph

TYPE OF FUEL
  PUMP . . . . electric

SOURCE OF FUEL . . . gas tank of servicer

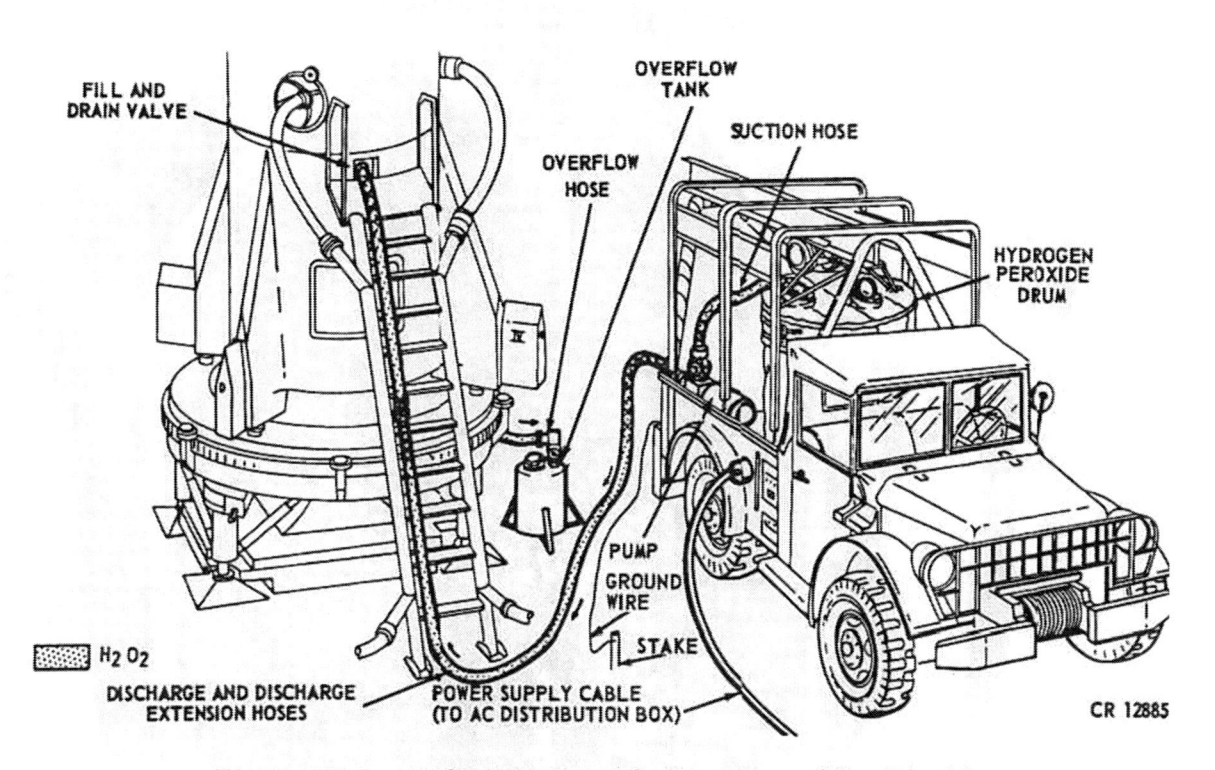

Figure VI-6 – Hydrogen Peroxide Servicing of the Missile

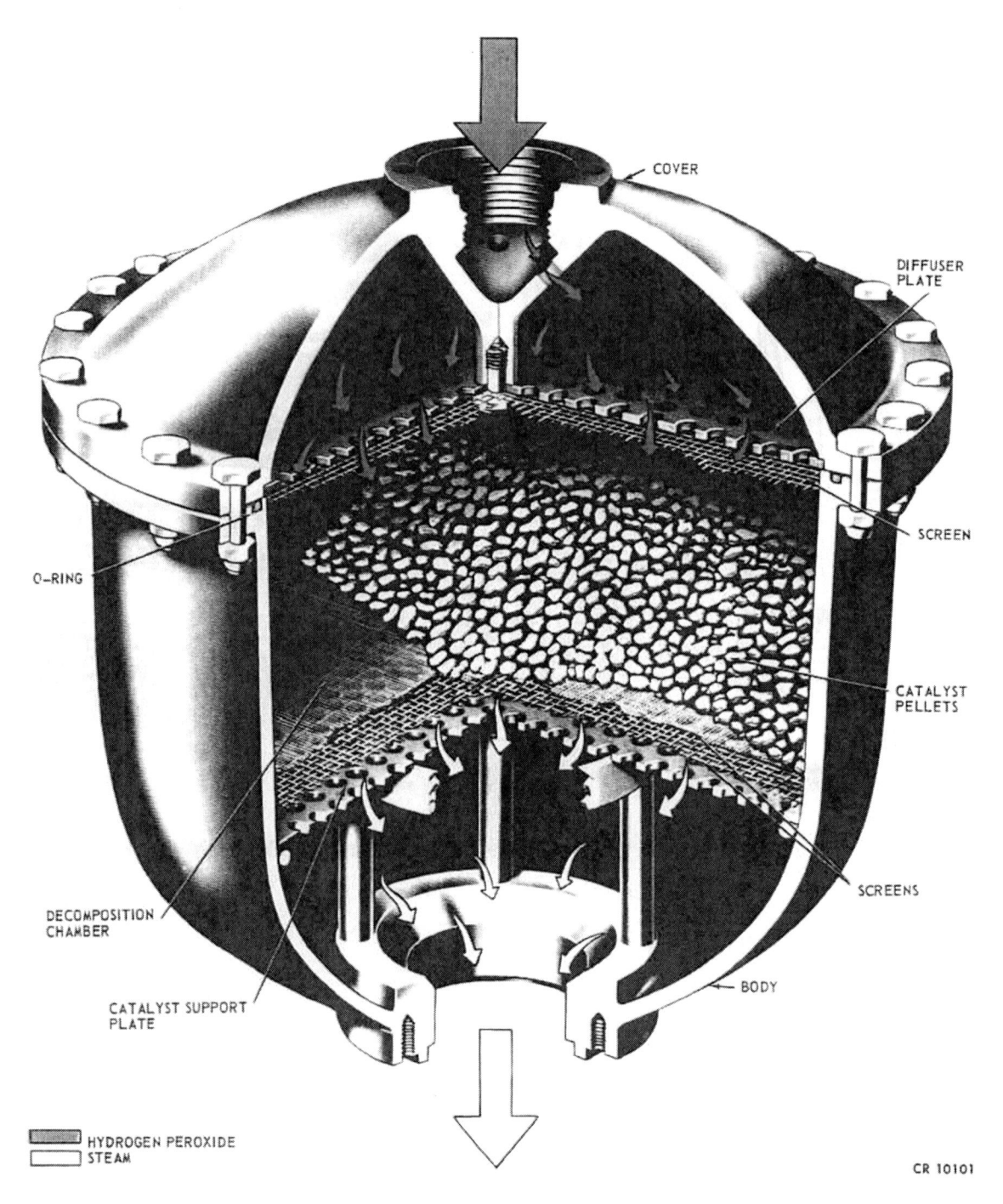

Figure VI-7 — Operation of the Steam Generator

VI-16

CHAPTER VII
PNEUMATIC SYSTEM

This page has been left blank intentionally.

CHAPTER VII

PNEUMATIC SYSTEM

In the REDSTONE Missile, the pneumatic system is used for pressurization, operation of valves, flight control, missile separation, and air bearings in the guidance system.

A supply of high-pressure air for the propulsion system and missile separation is carried in six spheres mounted in the missile tail unit. Also, two additional high-pressure spheres in the body unit supply air to the control system, air bearings, and separation system.

The air used in the missile system must be oil-free, dry air with a dew point of at least -65°F. Air pressure requirements range from a maximum of 3,000 psi to a minimum of 21 psi.

A Clark horizontally opposed, six-cylinder compressor unit driven by a gasoline engine supplies the air requirements for the missile pneumatic system. The compressor unit is equipped with a Vortox air filter which filters the air before it enters the first-stage cylinders and after it leaves the dehydrators. Moisture separator bottles are provided after each stage of compression for the purpose of condensing and removing water vapor from the compressed air. The dryer unit (or dehydrator) removes the remaining moisture and oil vapors. Purolator filters for removing foreign particles larger than ten microns in size from the compressed air are installed in the main air flow line after the oil and water dehydrators and before the spherical receiver. Oil-free, dry air at 5,000 psi is delivered from the air compressor to the air servicer.

The air servicer stores high-pressure air (5,000 psi) received from the air compressor truck and supplies regulated air pressure to the launcher valve box during missile pre-issue checkout, horizontal checkout, and vertical checkout. The air servicer also supplies high-pressure air (750 psi) to the LOX replenishing valve during the LOX replenishing operation. The air servicer chassis is a standard M-200, 2-wheel, heavy-duty chassis which has been modified by adding a rear bumper assembly and a subframe assembly on which the air servicer unit and floor assembly are mounted. The air servicer carries four air bottles which have a combined capacity of ten cubic feet. These bottles function as an air reservoir to assure a continuing air supply at periods of peak demand. In addition, they provide a reserve air supply when it becomes necessary to shut down the air compressor truck. Various valves, gages, regulators, and pneumatic lines are provided for adjusting, monitoring, and directing the path of pressurized air through the system.

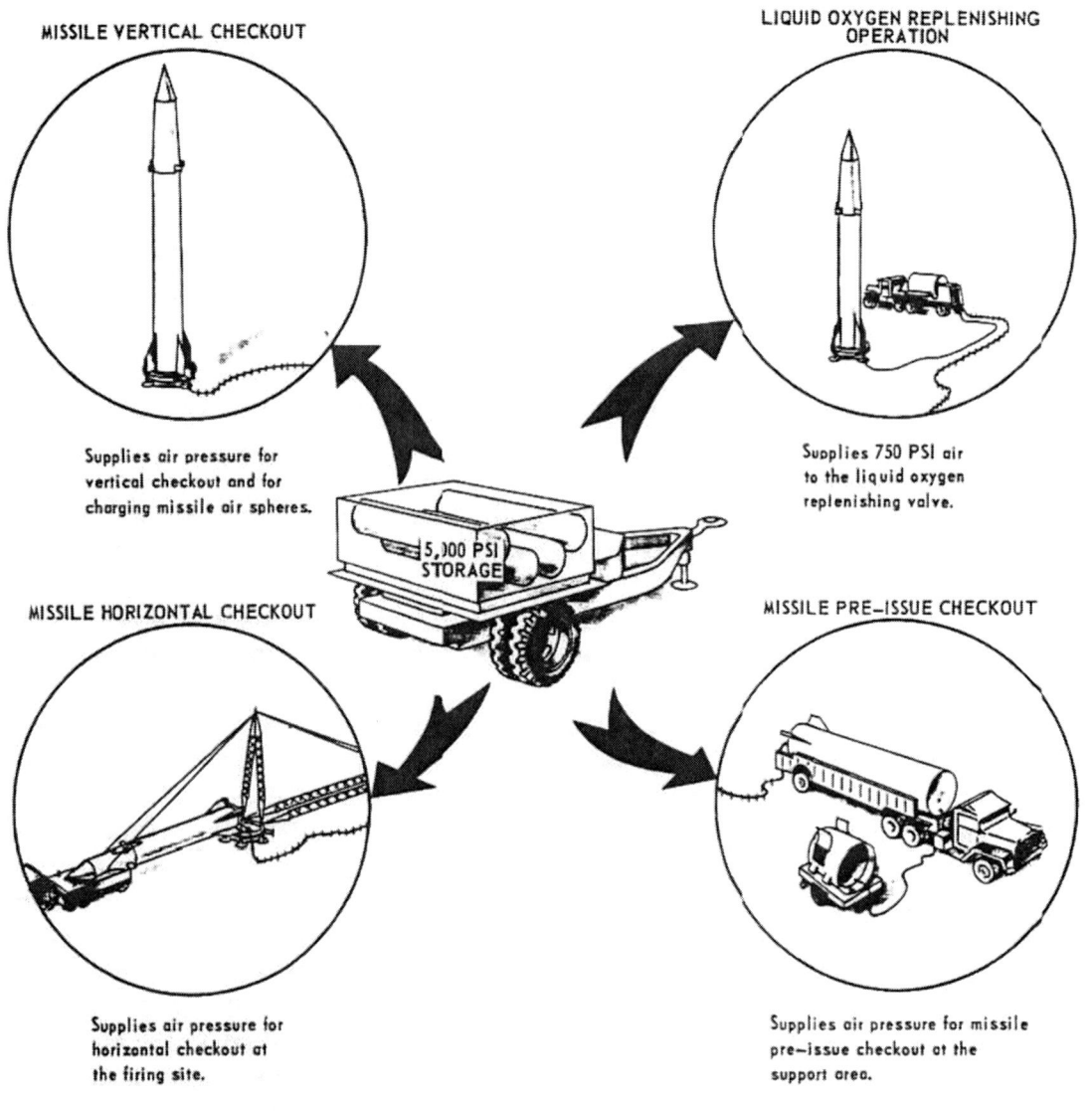

MISSILE VERTICAL CHECKOUT

Supplies air pressure for vertical checkout and for charging missile air spheres.

LIQUID OXYGEN REPLENISHING OPERATION

Supplies 750 PSI air to the liquid oxygen replenishing valve.

MISSILE HORIZONTAL CHECKOUT

Supplies air pressure for horizontal checkout at the firing site.

MISSILE PRE-ISSUE CHECKOUT

Supplies air pressure for missile pre-issue checkout at the support area.

CR 11941

Figure VII-1 — Function of the Air Servicer

High-pressure air (3,000 psi) is supplied to the ground control valve box by the air servicer until missile liftoff. All pneumatic connections between the missile and the valve box are made through a multiple coupling head on the valve box. The ground control valve box contains components such as valves, regulators, switches, and gages for control and testing of missile functions before and at the time of missile

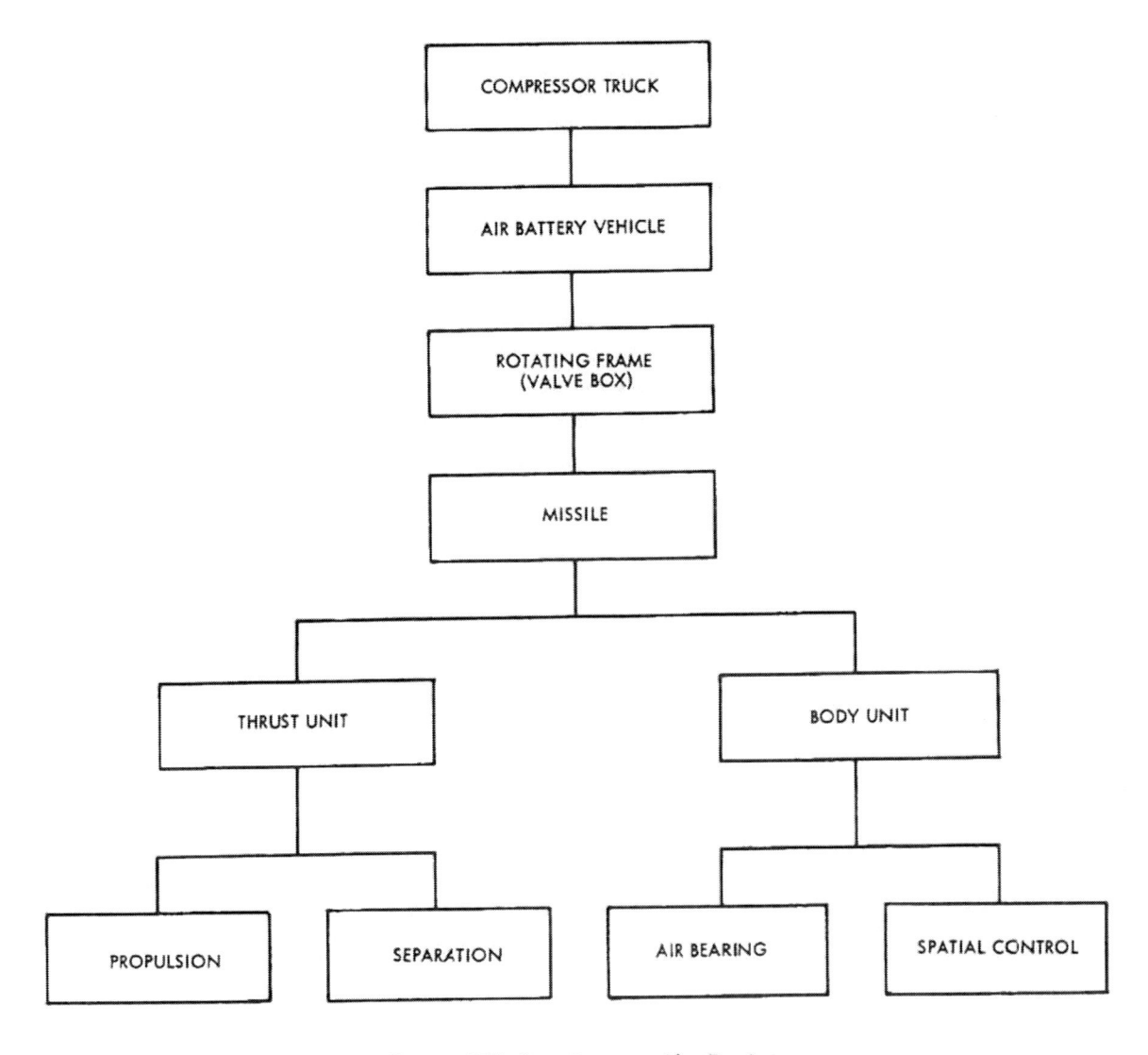

Figure VII-2 — Pneumatic System

firing. Normal control of the ground valve box is from the propulsion control panel (in the fire control and test truck) and from the remote firing panel (located in a foxhole).

A solenoid valve in the valve box allows air (3,000 psi) to pass through the multiple coupling into a high-pressure fill line. From a standard tee on the high-pressure fill line, the triplex spheres are filled to 3,000 psi.

From the triplex spheres, the air flows through a filter to the heat exchanger located in the steam exhaust duct from the turbine. The heat of the air is then increased to approximately 212°F. From the heat exchanger, air is fed through a cross, where it is routed to the top of the alcohol tank for pressurization (21 psi). Air also flows through the cross to the engine pressure regulator, where it is regulated to 585 psi by an internal loader. From here air flows to a four-way solenoid valve which controls the opening and closing of the main LOX valve and the main alcohol valve, and to a pneumatic manifold.

The manifold supplies air 1) for the main peroxide valve opening which is controlled by means of a solenoid valve, 2) for the peroxide tank pressurization by way of a shuttle valve, 3) to the alcohol tank vent valve through a solenoid control valve, and 4) to the expulsion cylinders to accomplish separation of the thrust unit from the body unit. A pressure relief valve is also mounted on the manifold as is an off-on pressure switch for confirmation of system pressurizing during the countdown sequence.

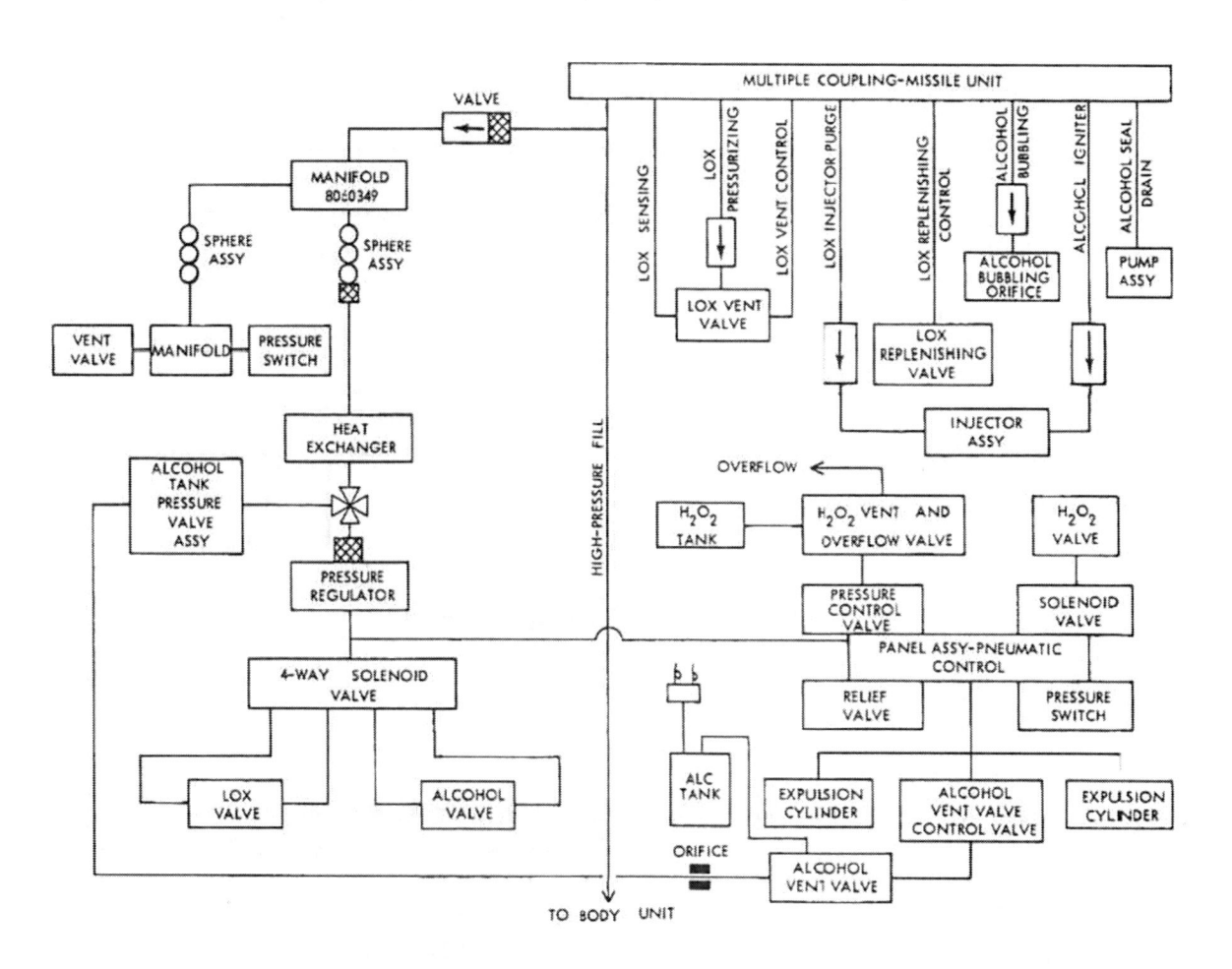

Figure VII-3 — Thrust Unit Pneumatic System

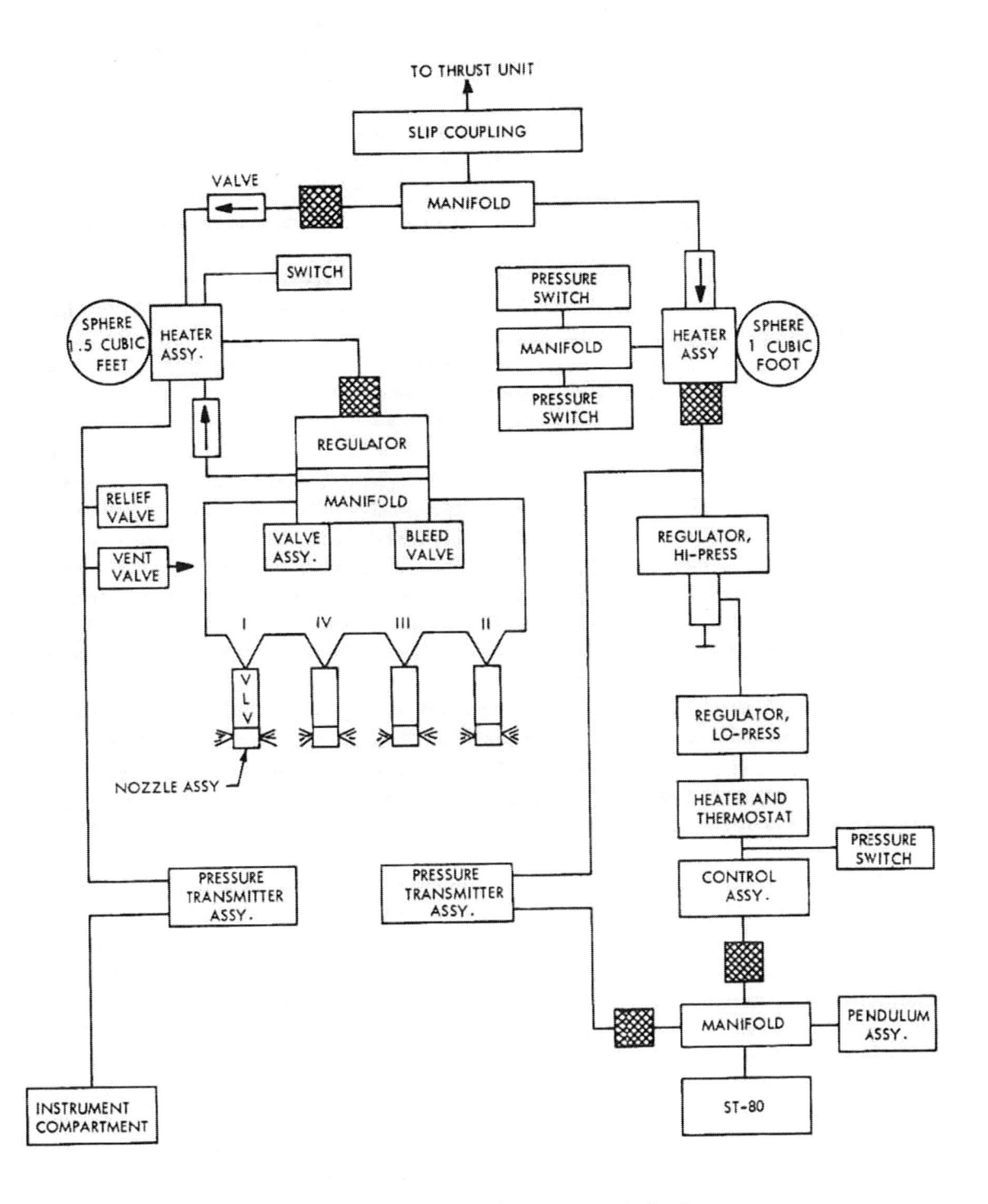

Figure VII-4 — Body Unit Pneumatic System

From the standard tee on the high-pressure fill line (to the triplex sphere assembly), a high-pressure line leads up to a slip coupling located at the top of the missile thrust unit. Mating with the coupling is an adapter in the missile aft unit and the air flow is directed to a manifold. From the manifold, separate lines run to the air bearing air supply sphere (1 cubic foot) and to the jet nozzle supply sphere (1.5 cubic feet). Pressure switches at the jet nozzle sphere, the air bearing sphere, and the triplex spheres are all actuated at 3,000 psi and are closed as part of the firing command circuit.

The jet nozzle system provides for spatial attitude control when the air vanes exceed an angle of ±5 degrees. Air, initially at 3,000 psi pressure, flows from the jet nozzle sphere through a filter to a pressure regulator. The pressure is reduced to approximately 300 psi by the regulator, which is adjusted by a loading device. From the regulator, the air then passes to a manifold and on to the jet nozzle solenoid valves. Commands to the electrical network system are derived from switches that are actuated by the movement of the air vanes. The electrical network, in turn, energizes the solenoid valves. Each jet nozzle is controlled by a separate solenoid valve. A relief valve to protect the downstream system and a bleed valve used during adjustment of the regulator are mounted on the manifold. A relief valve, a vent valve, and a check valve are attached to the heater assembly of the jet nozzle. The check valve is provided for unidirectional flow of bypassed air downstream of the pressure regulator back to the sphere. This flow will occur when the pressure in the sphere drops below that of the jet nozzle downstream of the pressure regulator. This pressure balance also protects the pressure regulator diaphragm. The pressure transmitter permits continuous supervision from the ground station of the pressure in the sphere and the instrument compartment.

An air supply and temperature control system is provided for the stabilized platform in the instrument compartment. When the stabilized platform is in operation (including test runs) prior to and after take-off, the pendulums, accelerometers, and stabilizing gyros are supplied with dried air, which is measured with a flow rate of approximately 11 scfm at 32.5 psi. This pressure will vary slightly depending on the individual ST-80 installation.

From the high-pressure sphere (1 cubic foot) the air flows through a filter to a pressure regulator which reduces the pressure to 65 ±5 psig. The air then flows past a gage connection and needle valve to a second pressure regulator, which further reduces the pressure. This second regulator is set to obtain 32.5 psig at the manifold. Air from the regulator flows through a heater and thermostat assembly to the orifice or bypass assembly. From the bypass assembly the air is passed through a filter to a manifold which directs the air to the ST-80 pendulum assembly and a pressure transmitter assembly.

The orifice or bypass assembly and pressure switch are included in the air bearing system to protect the air bearings in the event that, during checkout, the high-pressure air supply drops below 750 psig. When air pressure drops below 750 psig, the pressure switch causes a normally open solenoid valve in the bypass assembly to close, thus bypassing the air through an orifice. The reduced flow provides sufficient time for the gyros to run down before exhausting the air supply. The pressure switch also allows the stabilizing gyros to be started when pressure increases to the desired level (15 psig). Adjustment of the pressure regulator is done during ground checkout. Gage connections on the valve assembly and on the manifold are provided for the purpose of monitoring during adjustment. The pressure transmitter enables supervision of the pressures.

The LOX tank pressurization is controlled by a solenoid valve in the valve box which is actuated directly from the propulsion control panel and indirectly from the remote firing panel. The firing command switch on the remote firing panel initiates a series of operations, one of which is the LOX tank pressurization (to 31.5 psi).

Air at 3,000 psi from the high-pressure system is released through a hand-operated valve in the valve box to a pressure regulator. The regulator is adjusted so that 750 psi is released into the control valve system. A pressure gage indicates the pressure within the control valve system, and a safety valve prevents overpressurization. This 750-psi control system performs the following functions:

1) LOX tank venting is pneumatically controlled by a solenoid valve actuated from the propulsion control and remote firing panel. This valve is normally open. Air from the 750-psi system keeps the missile LOX vent valve open. When the solenoid valve is closed, pressure is cut off from the LOX tank vent valve and the vent valve closes.

2) The LOX replenishing fill and drain valve in the missile is pneumatically controlled by a solenoid valve also actuated from the propulsion control and remote firing panels. This solenoid valve is normally closed, and when opened, the pressure opens the normally closed LOX replenishing fill and drain valve located in the missile.

3) The injector plate is flushed with air in order to prevent the accumulation and freezing of moisture. A connection is made between the igniter alcohol pressurization fitting on the valve box and the injector purge fitting on the missile. Injector plate flushing is accomplished by operation of a hand valve.

4) After hose connections have been made between the valve box and the ignitor alcohol container, pressurization of the container is accomplished by opening a hand-operated valve. A solenoid valve controls the flow of igniter alcohol from the container to the missile during the ignition stage.

5) An air supply is provided by a solenoid valve for bubbling the alcohol at the alcohol pump inlet elbow of the rocket engine turbopump assembly. Bubbling begins during missile alcohol tank and LOX tank filling, and continues until liftoff. This bubbling prevents the possible freezing of alcohol in the pump suction line. Danger of freezing is caused by the proximity of LOX in the missile LOX tank over a projected length of time.

The LOX sensing system limits LOX tank pressurization. One of two pressure switches, which is used during testing, interrupts pressurization at 7 psig for leakage tests. At firing of the missile, pressurization is continued above 7 psig and is interrupted at 31.5 psig by the second pressure switch.

The leakage of the alcohol pump bearing seals is carried by a drain line which passes through the multiple coupling to the ground.

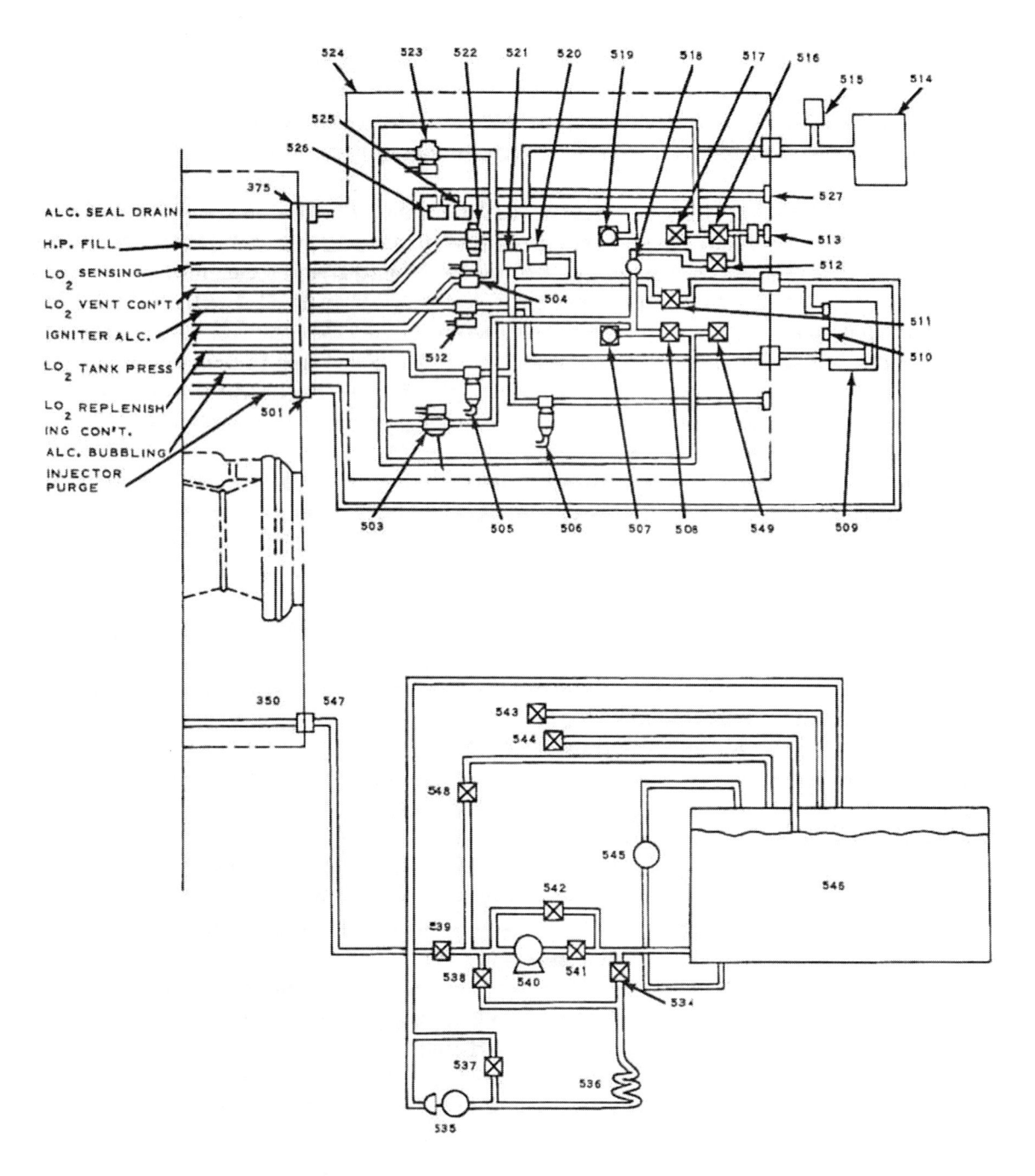

Figure VII-5 — Ground Control System

KEY TO COMPONENTS

350 — Slip Coupling, Missile Half
375 — Multiple Coupling Assembly, Missile Half

VII-8

501 – Multiple Coupling Assembly, Ground Half
502 – Igniter Alcohol Control Valve
503 – Pressurizing Valve, Solenoid-Controlled
504 – Pressurizing Valve, LOX Control
505 – Control Valve, LOX Replenishing
506 – Control Valve, Pressurizing
507 – Pressure Gage
508 – Valve, Alcohol Bubbling, Air Pressure, Hand-Operated
509 – Igniter Alcohol Container
510 – Igniter Alcohol Fill Inlet
511 – Pressurizing Valve, Hand-Operated
512 – Shutoff Valve, Filter, Hand-Operated
513 – Filter, High-Pressure Air
514 – Auxiliary Pressure Container
515 – Pressure Relief Valve
516 – By-pass Valve, Hand-Operated
517 – Pressure Test Valve, Missile Check Valves
518 – Regulator, High-Pressure Air
519 – Pressure Gage
520 – Pressure Switch, Alcohol Container
521 – Vent Valve, LOX Tank
522 – Control Valve, LOX Vent, Normally-Open
523 – Solenoid Valve, Missile High-Pressure Fill
524 – Ground Control Valve Box
525 – Pressure Switch, 7 psig
526 – Pressure Switch, LOX Control, 31 psig
527 – Gage Connection, LOX Tank Pressure-sensing
534 – Pump Suction Valve
535 – Regulator, Pressurizing
536 – Coil, Pressurizing
537 – By-pass Valve, Pressure Regulator, Hand-Operated
538 – Coil Inlet Valve
539 – LOX Trailer Coupling, Ground Half
540 – LOX Pump
541 – Pump Suction Valve
542 – Pump By-pass Valve
543 – LOX Vent Valve
544 – LOX Full Trycock Valve
545 – Pressure Gage, LOX Tank
546 – LOX Tank
547 – Slip Coupling, Ground Half
548 – Priming Line Valve
549 – Vent Valve, Alcohol Bubbling System, Hand-Operated

This page has been left blank intentionally.

CHAPTER VIII
FIRING SITE OPERATIONS

This page has been left blank intentionally.

CHAPTER VIII
FIRING SITE OPERATIONS

GENERAL

The purpose of this section is to provide a description of the operational sequence in which the REDSTONE Weapon System, including the missile and ground support equipment required at the firing site, might be deployed, prepared, tested, and fired during a tactical situation.

Because of the variables that are peculiar to each situation, the methods and procedures for conducting a firing operation, as described herein, should be regarded only as examples and not as established tactical concepts. The final decision concerning the approach, methods, and procedures necessary to the firing mission will be made by the Firing Battery commander, subject to the concurrence of his Group Commander.

Before the firing operations are covered, a brief presentation of the general organization of the Field Artillery REDSTONE Missile Group is included. The group can be considered the smallest self-sustaining element within the theater organization. The control center of the group is the Headquarters and Headquarters Battery, whose primary function is to direct and coordinate all group operations. The using units of the group are the two field artillery missile batteries (firing batteries). The support units consist of an Ordnance Company and Engineer Company.

The mission of the firing batteries is to serve as the firing components of the missile group, in general support of the field army, by projecting mass destructive fire on the enemy at ranges greater than those possible by conventional field artillery.

The primary function of the Ordnance Company is to provide ordnance maintenance and supply support for the missile and supporting Ordnance equipment. This company is responsible for the receipt, storage, inspection, acceptance testing, and issue of missiles, warheads, fuels, and line items which are not the direct responsibility of the Engineer Company. The Ordnance Company lends direct support to the Firing Battery by providing maintenance teams (with mobile field maintenance shops). These teams will move forward to the launching area when a malfunction occurs in a component or assembly for which the Ordnance Company is responsible and the troubleshooting and repair of which is not within the capability of the Firing Battery personnel and equipment.

The Engineer Company is responsible for three essential functions in its required support of the Field Artillery REDSTONE Missile Group in general and the Firing Battery in particular. One function is to generate, store, and transport LOX in sufficient quantities to meet the firing capability of the group. Another function is the generation, storage, and transportation of liquid nitrogen in sufficient quantities to meet group requirements. A third requirement of the Engineer Company is to provide field maintenance and repair on all basic motor-driven vehicles; that is, trucks, trailers, and diesel generators. The Engineer Company will supply contact teams in the launching area for the purposes of inspection, repair, or evacuation of items for which it is responsible.

All personnel of the Field Artillery REDSTONE Missile Group, with the exception of the chaplain and medical personnel, are equipped to fight as infantrymen, when required, to defend group installations against ground attack.

In the following discussion, it will be assumed that the firing order, with pertinent information, has been given to the Field Artillery REDSTONE Missile Group Commander by the Field Army Commander. This information would include: the geographic coordinates of the target, the desired time of missile firing, pertinent warhead information, and the general area of the launching site. It will further be assumed that a Corps of Engineers Survey Team has established two survey marks of at least third-order accuracy and separated by a distance of from 100 to 300 meters. The location of these two survey marks, their respective Universal Transverse Mercator (UTM) grid coordinates and elevations, and the grid azimuth between them will have been made known to the Survey Section of the Field Artillery REDSTONE Missile Group.

SITE SELECTION AND PREPARATION

Upon receipt of the firing order, the group commander will notify the respective commanders of a Firing Battery, the Ordnance Company, and the Engineer Company. The two support companies will immediately make preparations for the delivery to the Firing Battery of the missile and all other support equipment required for the firing operation when it is requested.

The Firing Battery Commander, accompanied by an operations officer and a survey team, will go forward to the designated general launching area for the purpose of selecting a suitable firing site. This reconnaissance may be conducted from the air or on the ground.

Because of the size and weight of equipment used to support a launching, the firing site selection is necessarily contingent on the following factors:

Accessibility

Roads, bridges, and trails leading to the firing site must be capable of supporting heavy equipment and must have sufficient width and clearance to permit passage of all the required loaded vehicles.

Size of Area

The size of the firing site is determined by the amount of space required to assemble and raise the missile and to accommodate the maximum concentration of ground support vehicles, which occurs during LOX filling. Within these limitations, over-all size will vary according to the ingenuity exercised in maneuvering vehicles and arranging equipment on the site.

Contour of Terrain

Uneven terrain can be used for a firing site provided the missile launcher can be leveled and the ground support equipment can be placed in an operational position.

Bearing Strength of Soil

The soil must have sufficient bearing capacity to support the emplacement of a launcher and a fueled missile under maximum wind load. The bearing capacity is determined as a figure called the cone index of the soil. In effect, the cone index is an empirical measure of the soil's shearing resistance and is determined with the aid of a cone penetrometer. The cone penetrometer is slowly pushed into the ground to a depth of 36 inches, if possible. Readings are taken at the surface and at 6-inch intervals. The average of these readings is the cone index. A cone index should be determined for each of the four launcher pad locations. The readings at these four locations should balance within 10 per cent. A cone index of 100 or greater is required for a single firing. This minimum figure provides a safety factor of one when the loaded missile is resting on the launcher and is under full wind load. If a multiple firing is anticipated from the same launcher location, the cone index must be 200 or greater.

Drainage

Natural and/or man-made drainage should be sufficient to prevent reduction of the soil bearing capacity around the launching area, especially around the launcher, in the event that the firing operation is undertaken during inclement weather. Drainage should also be sufficient to prevent standing water from hindering the firing operation in any way.

Cover and Protection

The firing site should have some natural cover to conceal the weapon system during the horizontal operations, which include firing site selection and preparation, vehicle deployment, and missile assembly and testing. Because of the height of the erected missile and because of LOX vapors and ice films, which can make the firing site prominent to enemy view and therefore vulnerable to attack, all vertical tests and firing preparations are completed as rapidly as technically feasible.

Survey Control

Because the azimuthal orientation of the missile is of paramount importance to the successful completion of the firing mission, and because this orientation is contingent, in part, upon the geographic location of the missile launcher, every effort is made to establish the firing site within a radius of 1000 meters of the surveyed marks which were described earlier. If this is not feasible, then the survey team should extend survey control from these survey marks to within 1000 meters of the launching point.

Lines of Sight

The firing site should afford clear lines of sight in order that the surveying and missile laying procedures can be performed rapidly, accurately, and with minimal opposition from natural obstructions.

Standby Area

An assembly area for the parking and dispersing of vehicles should be available near the launching site. Communication between these two points is necessary to assure the timely arrival of each vehicle as it is requested by the Firing Battery commander.

After all the aforementioned factors have been taken into consideration and an optimum suitability of the launching area has been realized, the area is ready for final preparation. This final preparation entails the clearance of trees, brush, and any debris which could hinder the vehicle march and emplacement, the surveying and missile laying, and/or missile assembly, testing, erection, fueling, and firing.

LAYING AND AIMING

The ultimate success of the firing mission is the delivery of the armed warhead into the desired target area within the prescribed radial tolerances.

The REDSTONE Missile is primarily a ballistic type, in that the greater part of its flight approximates the trajectory of a free-falling projectile. The entire flight is planned to follow a predetermined path from the launcher to the target. It becomes readily apparent that, to instill within the missile guidance system the necessary information to direct the missile to follow a fixed flight path, certain basic considerations are necessary. These considerations are: the geographic positions of the launching point and of the target with the resulting distance between them, their azimuthal bearing relationship, and the inherent errors of the composite missile.

The inherent errors of the missile are the result of all the imperfections, both material and operational, of the missile components and assemblies. The optimum condition of missile operation is therefore contained within the prescribed tolerances of the components and assemblies.

The firing azimuth that the missile must travel from the launching point to the target, and the distance between these points, are functions of the geographic coordinates and altitudes of these two points. Because the location and altitude of the target are known, the accuracy of the missile firing, irrespective of the composite missile tolerances, is dependent upon the accuracy in which the exact position and altitude of the launching point are determined.

PRELIMINARY CONSIDERATIONS AND PREPARATIONS

For every firing situation the individual pieces of laying equipment are restricted to emplacement either at fixed points or in a fixed general area. Because an optical method of laying the missile is employed, clear lines of vision between the pieces of laying equipment are essential. Any obstruction must be eliminated by the method most appropriate under the prevailing conditions.

The accuracy of optical equipment is affected by temperature variations. It is, therefore, necessary to protect this equipment from extremes in temperature changes. Umbrellas are utilized to protect the laying instruments from the direct rays of the sun or from the effects of rain, snow, or other adverse weather conditions.

The preliminary qualification adjustments to the laying equipment are the responsibility of the missile laying team and should be accomplished prior to arrival at the launch site. These adjustments are verified at the launch site prior to the performance of the laying procedures.

GIVEN INFORMATION

Certain information is required at the launching area in order to properly orient or aim the missile. As previously mentioned, the Corps of Engineers Survey Team is responsible for the establishment of two survey marks of at least third-order accuracy. The location of these two survey marks, their respective Universal Transverse Mercator grid coordinates and elevations, and the grid azimuth between them are made known to the Group Survey Section and the missile laying team. In addition to this information, the appropriate Transverse Mercator Projection tables and a map of the launching area sector are also needed to establish the coordinates and the elevation of the launching point.

> NOTE: The surface of the earth is divided into five areas as far as military mapping is concerned. Each area is mapped with respect to a selected spheroid and data which permit accurate representation of the mapped area. The Transverse Mercator Projection (TMP) represents a map of a sector on an appropriate mapping spheroid. The sector represented on the TMP is bounded by longitudes of 3 degrees East and 3 degrees West of a meridian and ranges from 80 degrees North latitude to 80 degrees South latitude.

EQUIPMENT UTILIZED

The following equipment is required in the launching area:

Two Wilde T-2 theodolites with night lighting equipment, theodolite mounted targets (TMT), tripods, and accessories.
One Wilde T-2 precise traverse target with night lighting equipment, tripod, and accessories.

One pocket transit (magnetic compass similar to an M-2 compass but graduated in degrees).

One 30-meter steel tape.

Two surveyor's umbrellas with supports.

Two flashlights with blackout filters.

Two nails (any size from 4 penny to 12 penny).

Recording forms and pencils. The forms are the Orientation Computation Sheet (OCS), 4 each, and the Commander's Verification Check Sheet (COVCS), 2 each.

PRELIMINARY LAYING

The determination of the coordinates and elevation of the launching point should be completed as rapidly as possible after the firing site has been selected and, preferably, before all the ground support equipment has been emplaced. The reason for haste is that the determined laying information must be submitted to the Group Firing Data Section for the development of final laying data.

As the laying equipment is brought into the launching area, the individual instruments are checked for verification of the accuracy of their preliminary qualification adjustments.

Of primary consideration in the laying operation is the proper positioning of the individual instruments. One of the Wilde T-2 theodolites, which is referred to as the Reference Instrument (RI), is established over one of the two reference survey marks. This survey mark is referred to as the Orienting Station (OS). The Wilde Precise Traverse Target (WPTT) is positioned over the other survey mark. This second survey mark is referred to as the Orienting Station Mark (OSM). The straight line joining the OS and OSM is referred to as the Orienting Line (OL). After the RI is plumbed over the OS and the WPTT is plumbed over the OSM, the RI operator sights along the OL to the WPTT. The azimuth scale of the RI is rotated to coincide with the azimuth reading of the OL, which was included in the required given information. A magnetic reading of the OL is taken by means of the pocket transit. The difference between the true heading and the magnetic heading is determined and recorded for later use during final laying.

With the aid of the aforementioned laying equipment, the Group Survey Section will determine the UTM grid coordinates and elevation of the launching point or launcher station (LS) by extending the existing survey control from the Corps of Engineers survey mark (OS) to the LS. For expediency, or of necessity, the missile laying team may be required to perform this standard survey operation. A survey stake, with a nail or some mark identifying the exact position of the LS, is then driven into the ground.

The results of the survey are transmitted to the Group Firing Data Computer Section in a manner most appropriate under the existing tactical circumstances. The computer section would normally be housed at or near the group headquarters. The computer section utilizes the LS survey information in conjunction with known target information to determine a geodetic azimuth or true line of fire (TLOF) at which the missile must be fired to strike the target. The computer section also provides the geodetic azimuth of the OL and specific guidance and control system presetting information.

Aside from laying equipment preparation and general emplacement, no further laying operations can be performed until the missile firing data have been received from the group computer section and the missile has been erected to the vertical position.

WEAPON SYSTEM EQUIPMENT REQUIRED AT THE FIRING SITE

The following is a list of major missile and ground support equipment required at the firing site:

Missile units and transport trailers
- (a) Warhead unit and semitrailer.
- (b) Aft unit and trailer.
- (c) Thrust unit and semitrailer.

Propellant service equipment
- (a) Alcohol tank semitrailer.
- (b) LOX tank semitrailer (2 each).
- (c) Hydrogen peroxide service truck.

Ground power and servicing equipment
- (a) 60-kw diesel generator (2 each).
- (b) Power distribution trailer.
- (c) Battery servicing trailer.

Missile launching and handling equipment
- (a) Erector-servicer truck.
- (b) Missile launcher platform.

Guided Missile Test Station

Pneumatic supply and storage equipment
- (a) Air compressor truck.
- (b) Air servicer trailer.

Stabilizer platform with shipping and storage container

Accessories transportation truck (2 each)

Fire fighting equipment
- (a) Fire truck.
- (b) Water tank trailer.

Liquid nitrogen container

LINE OF VEHICLE MARCH

The road march and vehicle emplacement is necessarily dependent upon the existing tactical situation and will be undertaken as directed by the Firing Battery commander.

The following is an example of an order of vehicle movement in a particular tactical situation:

1. The erector-servicer truck towing the Missile Launcher.
2. An accessories transportation truck towing a 60-kw diesel-generator trailer.
3. An accessories transportation truck towing the power distribution trailer.
4. The air compressor truck towing the air servicer trailer.
5. The stabilized platform shipping and storage container (including the ST-80) will be brought to the launch site by some vehicle designated by the Firing Battery commander.
6. The missile warhead unit (in its semitrailer) towed by a tractor truck.
7. The missile aft unit (in its trailer) towed by a cargo truck.
8. The missile thrust unit (in its semitrailer) towed by a tractor truck.
9. The liquid nitrogen supply vehicle.
10. The guided missile test station (GMTS) towing the battery servicing trailer.
11. The fire truck towing a water tank trailer.
12. The alcohol tank semitrailer towed by a tractor truck.
13. The first LOX tank semitrailer towed by a tractor truck.
14. The second LOX tank semitrailer towed by a tractor truck.
15. The hydrogen peroxide servicer truck.
16. An auxiliary 60-kw diesel-generator trailer towed by some vehicle designated by the Firing Battery commander.

EQUIPMENT EMPLACEMENT

After the launcher station (LS) has been established, and before the vehicles arrive at the firing site, a centerline should be established. This centerline must pass through the LS and extend between the intended positions of the erector-servicer truck and the warhead unit semitrailer. This centerline can be a chalk line or a light rope. If existing conditions permit, the centerline should be laid out in such a direction as to result in the erected missile being quasi-aligned on the desired target azimuth.

It may be necessary to use aircraft landing mats to support the launcher and other heavy equipment if the soil is softened or washed out because of rain or other adverse weather conditions.

The following procedure is a description of a proposed deployment of vehicles and equipment; tactical considerations may dictate variations:

1. The erector-servicer truck, towing the missile platform launcher, moves into the firing site and straddles the centerline. The launcher is positioned over the LS survey stake. The hydraulic cart is unloaded. The erector-servicer truck is positioned behind the launcher, over the centerline, and the A and H frames are unloaded and assembled.
2. A 60-kw diesel-generator, towed by an accessories truck, is positioned approximately 150 feet from the launcher. The accessories truck is unloaded.

3. The power distribution trailer, towed by an accessories truck, is centrally located near the 60-kw generator.
4. The air compressor truck, with its air servicer trailer, is positioned. The air compressor is started for a warmup.
5. Sufficient cabling should be completed to allow the 60-kw generator to be started with no necessity for shutting it down during later cabling exercises. The generator is then started for a warmup.
6. A public address system and/or other methods of inter-area communications are activated as rapidly as possible in order to provide a means of coordinating the total operation.
7. The stabilized platform, in its container, is brought into the area and conveniently positioned for preparation for installation in the missile.
8. The warhead unit semitrailer is positioned, straddling the centerline, approximately 150 feet ahead of the launcher. The cover is removed from the warhead unit trailer.
9. The aft unit trailer is moved, over the centerline, ahead of the launcher and under the A-frame chain hoist. The aft unit trailer cover is removed and the aft unit is lifted by the chain hoist.
10. The warhead unit trailer is backed up to the suspended aft unit and the two missile units are mated, still supported by the warhead unit trailer. The warhead unit trailer is then moved ahead with the complete body section.
11. The thrust unit semitrailer, with cover removed, is positioned over the centerline, just ahead of the launcher.
12. With the chain hoist, the rotating frame is lifted away from the launcher and attached to the base of the thrust unit.
13. The chain hoist is then used to lift the thrust unit from its trailer, and the trailer is removed from the immediate area.
14. The warhead unit trailer, containing the complete missile body section, is backed up to the suspended thrust unit. The body section and thrust unit are mated.
15. The warhead unit trailer is backed further until the rotating frame can be reattached to the launcher.
16. The chain hoist is removed from the thrust unit and the erecting cables from the A-frame are connected to the missile body section.
17. The liquid nitrogen supply vehicle is brought into the area and positioned approximately 45 feet from the launcher and in the sector that conforms to the position of the installed heater-cooler drop tank.
18. The battery service trailer, towed by the Guided Missile Test Station (GMTS), is positioned near the 60-kw diesel-generator. The GMTS is then positioned approximately 25 feet from, and to the side of, the launcher.
19. Battery activation is started as quickly as possible.
20. The fire truck and water tank trailer are positioned in such a manner as to be readily available to combat any fires that might occur in any area.

21. The propellant service vehicles (the alcohol trailer, the two LOX trailers, and the hydrogen peroxide truck) are parked apart from each other and at a considerable distance from the firing site. This is a safety precaution because of the volatility of these substances. This parking method also reduces the congestion in the immediate area of the firing site.

The sand-bagged emplacement that houses the remote firing box is set up as the Firing Command Post. This emplacement should be positioned in such a manner as to permit an unobstructed view of the erected missile and the firing site in general, while affording sufficient protection for the personnel manning the post. The emplacement can be made at a distance of up to 200 yards from the launcher.

With the positioning of the propellant servicing vehicles, the initial vehicle emplacement is completed, and the stage is set for the preparation for the horizontal checkout of the missile.

HORIZONTAL CHECKOUT

This phase of the firing mission is directed toward the functional testing of the components and assemblies of the weapon system. In other words, the operational readiness ("go - no go" condition) of the complete weapons system is ascertained during this phase.

Preparation

Before the various tests can be performed, it is necessary that the electrical and pneumatic interconnections required between the missile and ground support equipment have been made. Ground power must be made available to all required areas. Qualified personnel must be on hand to perform the necessary operations. There must be a complete system of intercommunications throughout the entire firing site complex. The stabilized platform (ST-80) is installed in the instrument compartment of the body section, but the platform is not electrically connected.

Tests Performed

Many of the operations performed during the horizontal checkout phase will be carried out simultaneously. For simplicity, the tests performed will be discussed individually and in a feasible order as follows:

Pneumatic System Checkout — The dewpoint of the high-pressure air in the air servicer trailer is determined. After the dewpoint has been determined and all air lines have been connected, the air is admitted into the ground and missile pneumatic system. The entire pneumatic system is checked for leaks, and the various valves and regulators are tested and adjusted as required.

Horizontal Power Check — After the electrical cabling has progressed to a point which permits the application of power to the missile system, tests are undertaken to assure that the various power sources, both ground and missile, are properly connected into the system and are operating within prescribed tolerances.

Sequence Recorder Check

Powerplant Components Test — This test involves the checkout of all electrically and pneumatically controlled valves which appear in the missile propellant system.

Range Guidance Computer Test

Lateral Guidance Computer Test

Warhead Arming Check — This is a check of the velocity and displacement arming switches located in the guidance computers. These switches set up the parameters within which the missile warhead can be armed.

Control System Test — This is a checkout of all the circuitry which maintains the proper attitude of the missile during its flight.

Inverter Calibration Test — Because of the critical areas in which the missile inverter power is used, it is necessary that the inverter's output (both the frequency and the voltage) be maintained within very critical tolerances.

Thrust Controller Test — The thrust controller maintains the combustion chamber pressure at a constant value during the propulsion phase by indirectly controlling the flow rate of the propellants.

Program Device Test

Over-all Cutoff Tests — These tests provide a method of checking the circuitry involved in the automatic shutdown of the missile system under various abnormal circumstances.

Horizontal Simulated Flight Test — This test serves as a check of the composite missile system throughout the various phases from the issuing of the firing command to impact. The functions that cannot actually be performed, for example, propulsion ignition, pyrotechnic device firing, and missile separation, are simulated to effect a sequential test of the over-all system.

Missile Batteries Activation and Installation — The missile system requires three batteries (two 28-volt and one 60-volt) during actual flight conditions. Ground power is used to simulate missile battery power during component and horizontal simulated flight tests.

Power Transfer Test — The purpose of this test is to check the operation of the missile batteries under normal load conditions. The power transfer test is the last of the horizontal tests to be performed. With its completion, the missile is readied for erection and vertical testing.

This phase is directed toward the final prefiring check to assure that the missile system is operationally ready, within the prescribed tolerances, to fulfill the firing mission.

Preparation

All test cables, regulator pressure gages, and test equipment peculiar to the horizontal checkout phase are removed and stored. This includes such test equipment as the stabilization system amplifier load box, the booster servo interrupter box, and the simulated flight test box.

The stabilized platform (ST-80) is electrically connected to its amplifier boxes, the instrument compartment access doors are sealed, and a leak check is performed to verify the pressure seal of the instrument compartment.

The heater-cooler drop tank is prepared and installed on the missile. The launcher is leveled, and the missile is erected with the aid of the A-frame. After the missile is placed in a vertical position on the launcher, the H-frame is assembled to provide a platform for personnel who are servicing the vertical missile.

All necessary interconnections, both electrical and pneumatic, are made and verified.

The air rudders and carbon jet vanes are inspected and installed.

The missile is rotated for coarse target azimuth alignment.

Tests Performed

The number of tests performed, and the order in which they are undertaken, may vary with each tactical situation.

Vertical Power Check – This is a final test to assure that the system power sources are properly connected and operating.

Preset Timers Check – Presetting adjustments in the cutoff computer are very critical and, as the accuracy of presetting is dependent upon timing, care must be taken to assure against intolerable error in the timing device.

Stabilized Platform Presetting – The firing mission data sheet, provided by the Group Computer Section, contains certain presetting information for the stabilized platform (ST-80). This information includes the elevation angle (epsilon angle) of the range accelerometer and the earth-rotation bias settings for the three missile axes.

Control System Test – This is a final test of the guidance and control system components which are responsible for maintaining the attitude of the missile during the flight phases.

Vertical Simulated Flight Test – This is a check of the composite missile system from the firing command to 20 seconds after simulated liftoff. This provides a functional check of most of the circuitry involved in the propulsion system as well as the power transfer sequence and limited operation of some of the guidance and control components.

FINAL LAYING

Final laying should be started as soon as possible after missile erection, upon receipt of firing data from the Group Computer Section.

When the geodetic azimuth of the true line of fire (TLOF) has been made known from the firing mission data sheet, the missile laying team can position the second theodolite, which is called the azimuth control instrument (ACI). The ACI is responsible for the final orientation of the missile to the TLOF. An optical reflecting instrument, called a porro prism, is located on the ST-80 mounting frame in the missile between missile body fins I and IV and just slightly forward of fin IV. The face of the porro prism is visible from outside of the missile in a visibility-restricted sector extending away from the general area of missile fin IV. The porro prism is physically oriented so that its dihedral edge is parallel to the vertical plane passing through fins I and III and at the same time, parallel to the local horizontal (assuming the missile has been vertically leveled). The final laying of the missile will have been accomplished when the vertical plane passing through fins I and III is aligned along the azimuth of the TLOF.

The method used in the orientation of the missile is referred to as the perpendicular sighting method. In order for the ACI operator to receive an optical reflection from the porro prism, he must be sighting in a plane that is exactly perpendicular to the dihedral edge of the porro prism. From this it can readily be seen that to optically align the missile to the TLOF, it will be necessary to rotate the missile until the dihedral edge of the porro prism is parallel to the TLOF.

The final laying exercise is described in the following paragraphs.

Determination of the Magnetic Azimuth of the TLOF

The magnetic azimuth of the TLOF is determined by adding to or subtracting from (as is necessary) the geodetic azimuth the azimuthal difference between the true bearing and the magnetic bearing at the particular point on the earth's surface on which the launcher station (LS) is located. This azimuthal difference was determined and recorded during the preliminary laying exercise.

Determination of the ACI Sighting Magnetic Azimuth

The ACI sighting magnetic azimuth is determined by subtracting 90 degrees (or adding 270 degrees) to the magnetic azimuth of the TLOF.

Determination of the First ACI Ground Mark

A laying team specialist, standing near the booster air rudder IV and utilizing the pocket transit, sights along the ACI sighting magnetic azimuth. While maintaining the sighting along this azimuth, the specialist moves directly backward and away from the launcher to a distance of 30 to 60 meters. At a suitable spot on this line just traversed, the specialist will make the first ACI ground mark.

Determination of the Distance Between the ACI and Launcher

By means of the meter measuring tape the specialist measures the distance from the ground mark to the inside ring of the launcher to the nearest decimeter.

NOTE: The inside ring of the launcher is used as a terminus because it is roughly in the same perpendicular plane to the local horizontal as the face of the alignment porro prism.

Qualification of the ACI Azimuth Scale

The reference instrument theodolite (RI), which had previously been positioned and qualified over the orienting station (OS) during the preliminary laying exercise, is turned in azimuth to sight on the theodolite-mounted target (TMT) mounted on the ACI. The ACI is turned to sight on the TMT on the RI. The azimuth scale reading of the ACI should be the back azimuth of the azimuth scale reading of the qualified RI. The azimuth scale of the ACI is rotated to read the azimuth reading of the RI +180 degrees (or -180 degrees, as is necessary). The ACI is then considered qualified in its azimuth scale.

Missile Coarse Alignment

The missile is rotated so that the vertical plane that passes through vanes II and IV is roughly parallel to the ACI sighting magnetic azimuth, and the porro prism is sighted by the ACI operator.

The missile and the ACI are alternately rotated until autoreflection is achieved. This first ACI azimuth scale reading is recorded.

NOTE: This reading undoubtedly will be other than the desired reading (geodetic azimuth of TLOF -90 degrees or +270 degrees, as is necessary), and further operations must be undertaken.

Determination of the Reposition Constant

To arrive at the desired firing azimuth, it will be necessary to turn the missile and/or reposition the ACI. The missile can be turned easily. All that is necessary is to determine how much and in what direction the ACI must be displaced in order to achieve missile target alignment.

Because the firing azimuth is critical to within minutes of arc, it is necessary to position the missile within a minute of arc. The SIN-TAN of this minute angle is 0.0002909, or in round figures, 0.0003.

The recorded distance (D) from the ACI ground mark to the launcher (measured to the nearest decimeter) is converted into centimeters.

VIII-14

The direction that the ACI must be displaced is easily ascertained by comparing the actual azimuth scale reading of the ACI and the desired reading, TLOF -90 degrees (or +270 degrees, as is necessary).

The amount of necessary displacement of the ACI can be computed as follows:

d (amount of necessary displacement) = D tan (or sin) Θ

Θ is the difference between the aforementioned angles.

0.0003 D (cm) = cm/min of arc = Reposition Constant (Rk)

Repositioning of the ACI

After determining the angle (Θ) and the amount and direction of necessary ACI movement, the laying specialist accurately measures the distance in the proper perpendicular direction from the first ACI ground mark and makes a second ground mark. The ACI is plumbed over this second ground mark and its azimuth scale is requalified by again bucking (cross-sighting) with the RI.

Missile Fine Alignment

The missile and the ACI are again alternately rotated until auto-reflection is achieved. It is possible that the missile still is not correctly oriented, and further computation, ACI repositioning, and missile rotation are necessary. Each time the ACI is repositioned, it must be requalified in order to maintain its degree of accuracy.

The method of final laying, with the necessity for repositioning the ACI, is a trial-and-error method. However, an experienced laying specialist can perform the operation quite rapidly and with a minimum of ACI movement.

FINAL PREPARATIONS FOR FIRING

When all the proposed tests have been completed to the satisfaction of the Firing Battery commander, the final firing preparations are begun.

Preparation for Propellant Loading

Some of the fueling preparations can be started as soon as the missile has been erected. This activity would include the connection of the various alcohol and LOX lines and valves and the positioning of the fueling ladder. The fire-fighting equipment should be in the area during propellant loading operations.

Inert Lead Start Fluid Filling

The powerplant (engine) jacket is filled with lithium chloride, which is carried in a container on the alcohol trailer, to permit a smooth transitional flow of alcohol during the engine starting phases.

Alcohol Filling

This operation is started after the vertical simulated flight test has been performed. The alcohol trailer is positioned approximately 20 feet from the fueling ladder. The igniter alcohol container, located on the launcher, is filled by gravity flow from the alcohol trailer emergency valve. The alcohol trailer is connected, through the fueling ladder, to the missile alcohol tank, and the alcohol is pumped into the tank. When the missile alcohol tank is full, the alcohol trailer is disconnected and driven out of the immediate launching area.

LOX Filling

After the alcohol trailer has left the immediate area, the two LOX trailers are moved into position. They are simultaneously connected, through a "Y" connection, to the fueling ladder and then to the missile LOX tank. LOX pumping is started from one of the LOX trailers. After a delay of three or four minutes, the pumping of the second LOX trailer (replenishing trailer) is started. This delay is to assure that the depletion of LOX in the second trailer will be less than that of the first, after LOX filling is completed. When the missile LOX tank is full, the LOX pumping operation is stopped and the first LOX trailer is disconnected and driven out of the immediate launching area. The second LOX trailer (the replenishing trailer) is disconnected and moved away from the launcher to a distance of approximately 150 feet. A LOX replenishing line is then connected from the replenishing trailer to the replenishing arm, mounted on the launcher and connected to the missile LOX tank. Periodically, throughout the rest of the operation until missile firing, the normal boiloff of LOX from the missile LOX tank will be replaced by the operator of the remote firing box. A replenishing switch controls the LOX replenishing.

Hydrogen Peroxide Filling

After the LOX filling exercise is completed, the hydrogen peroxide truck is positioned near the missile, and the container is connected to the missile hydrogen peroxide tank. The hydrogen peroxide is then pumped into the missile tank. The hydrogen peroxide, in conjunction with catalytic pellets, produces the steam which drives the propellants turbopump.

Upon completion of the hydrogen peroxide filling, the hydrogen peroxide truck is disconnected and driven from the immediate area.

Vertical Range Computer Test and Presetting

This exercise includes the final check of the operational status of the range guidance computer and the cutoff computer, which are housed in the same assembly.

Upon verification of the in-tolerance operation of the range computer, the presetting information listed on the firing mission data sheet is instilled into the cutoff computer. This presetting information includes constant values of velocity and displacement, which are compensation for inherent errors peculiar to the individual trajectory of the missile.

Vertical Lateral Computer Test

This test is designed to effect a final check of the operational status of the lateral guidance computer.

Warhead Prelaunch Check

This check determines the condition of missile network circuitry related to the warhead. The type of warhead burst (air or surface) is also selected during this exercise.

Final System Preparations and Equipment Removal

The guided missile test station (GMTS) is disconnected from the missile system and driven out of the firing area.

The powerplant ignition system is readied and the blind plugs, which serve as deactivating or safety devices, are installed.

The missile laying team makes a last check to assure that the missile is properly oriented to the TLOF.

The immediate firing area is cleared of extraneous equipment and personnel.

The remote firing box operator operates the LOX replenishing switch until LOX tank overflow is visible.

At the predetermined time and on order from the Firing Battery commander, the remote firing box operator closes the firing switch.

MISSILE FIRING

The general sequence of operation of the missile system after the firing command given, is as follows:

1.) The missile propellant system vent valves, which were opened for the propellant filling exercises, are closed.

2.) The missile pneumatic system is then utilized to pressurize the propellant tanks. The pressurization of the hydrogen peroxide and alcohol tanks is started immediately upon the firing command signal. When the alcohol tank pressure is sufficient, a pressure-sensing switch actuates to stop alcohol tank pressurization and to start the LOX tank pressurization.

NOTE: Throughout a portion of the flight, the propulsion system tanks pressurization is cycling because of the action of pressure-sensing switches.

3.) When the LOX tank pressure has reached a specific amount, a pressure switch is actuated to stop the LOX tank pressurization and to start the power transfer sequence.

NOTE: The missile system has been operating from ground power until this time in order to save the missile batteries. These batteries are very short-lived under normal loading conditions.

During the power transfer sequence, the missile batteries are connected to their respective missile busses, in parallel operation with the ground power sources. When the tie-in has been accomplished, the ground power sources are disconnected from the missile system. The operation results in a smooth transition from ground power to missile battery power.

4.) The completion of the power transfer sequence initiates action to propel the heater-cooler drop tank away from the missile.

5.) The physical breaking of the electrical connections between the missile and the heater-cooler drop tank actuates the ignition starting sequence. After a slight time delay, which allows the three missile batteries to settle down under the rather rapid loading action of the power transfer sequence, the igniter squib is fired and ignition takes place. LOX, in the ignition stage, is taken directly from the missile LOX tank under gravity flow while the alcohol is obtained under pressure from the igniter alcohol container located on the launcher.

6.) The mainstage stick link is physically located so as to be burned through by the heat of combustion within the engine. This results in the opening of the main alcohol and the hydrogen peroxide flow valves in anticipation of thrust buildup. The hydrogen peroxide flows onto the catalytic pellets in the steam generator and causes a steam buildup which drives the turbopump. The turbopump, in turn, drives missile alcohol and LOX into the combustion chamber, where the mixture is burned to produce engine thrust.

7.) When the engine thrust has developed to the point of overcoming the gravitational effects exerted on the missile mass, the missile rises from the launcher. At this point, the flight phases are started and the missile guidance and control system is activated.

8.) The program device is actuated to begin triggering certain critical electrical circuitry.

9.) The stabilized platform (ST-80) loses its earth-fixed reference and becomes inertially stabilized in space. Any attitude deviation of the missile from the prescribed trajectory is detected by the ST-80 and, through the operation of the missile control system, the deviation is cancelled out.

10.) The guidance computers are activated to detect and store any translational deviations of the missile from the desired trajectory.

11.) The cutoff computer is started and monitors the speed of the missile. When the missile reaches a predetermined velocity, this computer will shut down the propulsion system by cutting off the propellant flow to the engine. The missile will then be in the ballistic, or free-falling, phase of its flight.

12.) The propulsion system shutdown is followed by thrust decay, or burnout of the propellants remaining in the propulsion system downstream of the main propellant valves.

13.) At a predetermined time, the pulse step switch actuates the separation circuitry, and the missile thrust unit is forcibly detached from the body section.

14.) When the body section comes back into the sensible atmosphere, a "Q" switch is actuated. This is the start of the re-entry phase and the end of the ballistic part of the missile flight. The actuation of the "Q" switch also signals the start of the terminal guidance operation. During the flight, the guidance computers had detected and stored the translational deviations of the missile from the prescribed trajectory. After re-entry, this information is fed to control system in order to direct the missile back toward the prescribed trajectory to assure that the warhead will land within the radial tolerance distance of the target.

RETESTING, ABORT FIRING, AND POST-FIRING OPERATIONS

Retesting

If a misfire occurs after the spring-loaded fire switch on the remote firing box has been operated, the normal procedure is to wait 30 seconds and then push the fire switch again. If the missile has not fired within 1 minute after the initial command, the emergency cutoff switch is actuated. A misfire constitutes a condition that requires retesting. Any extended interruption between the time of guidance presetting and the firing command also necessitates retesting.

In the event that retesting is necessary, certain precautions should be taken and procedures followed to assure the safety of both personnel and equipment.

1.) The missile pneumatic system is vented, from the remote firing box, for 3 minutes.
2.) The blind plugs are removed to disable certain critical circuits.
3.) All necessary personnel and equipment start forward to the immediate launching area.
4.) The GMTS is electrically connected to the missile system in a definite sequence in order to prevent any circuit activations or interruptions.
5.) The missile control system is disabled from the GMTS, and the missile system is in a state which will permit either retesting or shutdown.

Retesting entails the troubleshooting of the weapons system to determine the reason for the misfire, the rechecking of the operation of the guidance and control components, the rechecking of the missile-target (laying) orientation, and the presetting operations.
NOTE: It may be necessary to circulate and heat the alcohol if the hold is long enough to drop the alcohol temperature below a specific limit.

Abort Firing

In the event that the firing mission cannot be carried out after all preparations have been completed up to the firing phase (because of equipment failure or orders cancelling the mission), the missile must be de-activated and disassembled in a specific sequence.

NOTE: The propellant tanks vent valves should be kept open during all abnormal downtime. Also, the fire-fighting equipment should be present in the immediate area.

The LOX trailers are reconnected, through the fueling ladder, and the LOX is transferred back to the trailers by gravity flow. The LOX trailers are disconnected and driven away.

The alcohol trailer is reconnected, through the fueling ladder, and the alcohol is transferred back to the trailer by pump action.

The hydrogen peroxide truck is reconnected to the missile hydrogen peroxide container, and the hydrogen peroxide is transferred back to the truck container by pump action.

All pyrotechnic devices, rudders, and carbon vanes are removed and stored.

All cabling and hoses are disconnected and stored.

With the aid of the erector vehicle, the missile is lowered, disassembled, and reloaded upon the transport trailers.

All equipment is stored and the road march of the vehicles is started.

Post-Firing Operations

If the missile firing was successful, there remains only the shutting down of the ground power equipment and the disconnection and storing of all equipment in preparation for leaving the launching area.

CHAPTER IX
TELEMETRY SYSTEM

This page has been left blank intentionally.

CHAPTER IX
TELEMETRY SYSTEM

GENERAL

Telemetry, by definition, is the complete measuring, transmitting, and receiving apparatus for indicating, recording, or integrating at a distance, by electrical translating means, the value of a quantity.

Since guided missiles, generally speaking, are not recovered after launching, telemetry provides a means of measuring physical quantities within the missile after launch. The physical quantities measured may vary from stresses on the vehicle skin to minute voltage changes in the power supplies. The number of measurements that can be made at one time varies with the type of measurement and system limitations. More than one telemetry package can be used if it is necessary to expand the amount of instrumentation on a missile.

The data transmitted via telemetry from a missile are received at a ground station, and the composite data are recorded. The recording can later be played back and individual data removed from the composite signal for processing and reduction.

TELEMETRY STANDARDS

Frequency band: 216 mc to 235 mc
  NOTE: Band may extend to 265 mc
RF Bandwidth: ±150kc

Eighteen different channel frequencies (subcarrier frequencies) can be applied to modulate the rf carrier. The 18 channels may be applied to the carrier either individually or simultaneously.

Each of the 18 channels may carry more than one bit of instrumented data by time-sharing or by commutating the channel.

Channel frequencies, deviation, and upper and lower frequency limits are shown in table IX-1.

TYPICAL TRANSMITTER OPERATION

The transmitter is of a PAM-FM-FM type. Refer to figure IX-1. This means that the sensing device has a pulse output of varying amplitude determined by the quantity being measured. The pulse amplitude, in turn, is frequency-modulating the subcarrier oscillator. The subcarrier oscillator output is multiplexed with other channel outputs to frequency-modulate the transmitter. When a phase-modulated transmitter is used the system is a PAM-FM-PM type.

Commutation of several inputs is done only when the quantity being measured is not so critical that it has to be constantly transmitted, or it is a slowly-changing quantity. A commutated channel also has to be synchronized with the ground decommutator for data processing.

A calibrator is used to periodically feed the channels with fixed reference-level signals for data reduction. The calibration is programed on and off within the missile in order to assure that the calibrator will be off at a time of some critical measurement such as cutoff or separation of the missile.

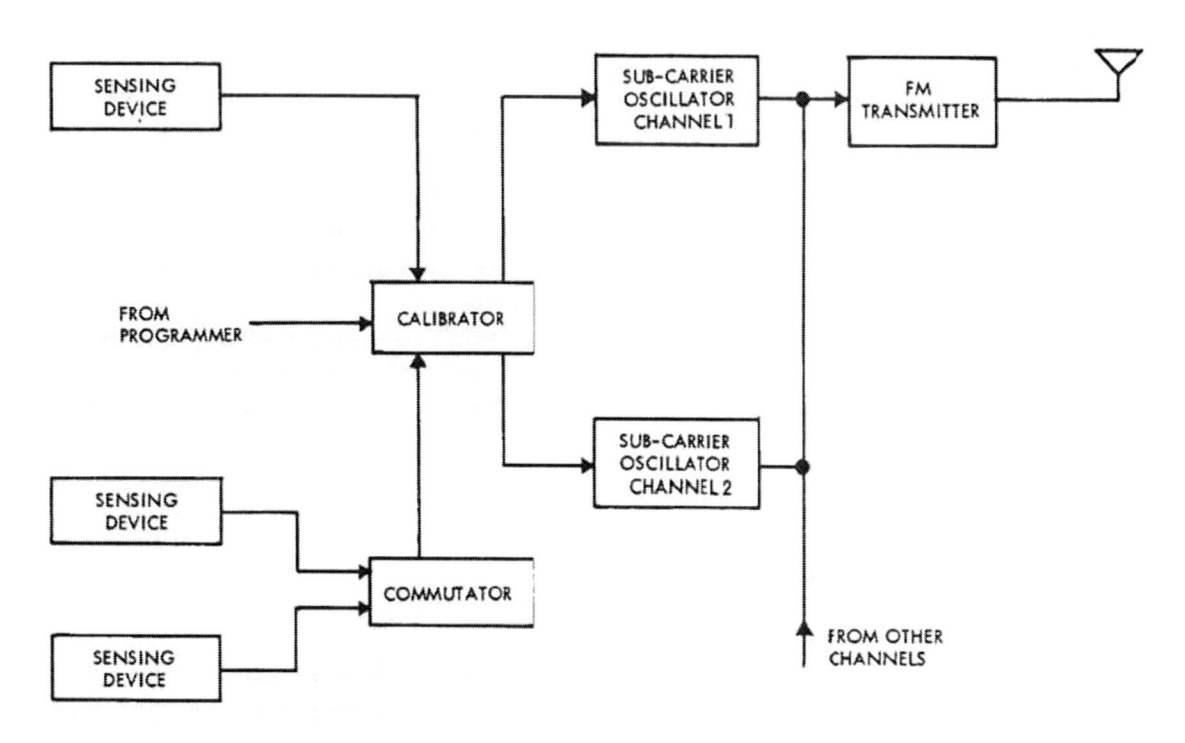

Figure IX-1 – Transmitter - Simplified Block Diagram

TELEMETRIC DATA TRANSMITTING SYSTEM AN/DKT-8 (XO-2)

This FM-FM system has a nominal power output of 35 watts and has a capacity of 17 subcarrier oscillators. Channel one is not used. Commutation is provided on one channel at a rate of 10 revolutions per second (rps). Information inputs to the straight channels may be either dc voltage type signals in the 0- to 5-volt range or frequency-type inputs such as that provided by a vibroton gage or external subcarrier oscillator. In the latter case, a band-pass filter is substituted for the subcarrier oscillator. Thus, the total system capacity is 16 straight plus 27 commutated inputs, or a total of 43 information inputs.

In some cases, the output of a sensing device is not of a voltage nature or is a very low voltage. In either case, an adapter is used to convert the output to representative voltages between 0 and 5 volts dc to modulate the subcarrier oscillator.

The telemetry package consists of two main assemblies, the main telemetry assembly and the power amplifier assembly.

The main telemetry assembly consists of:

Subcarrier oscillators (560 to 70,000 cps).
Distribution panel.
Video amplifier.
Transmitter.
150- to 250-volt power supply.
Commutator gating unit.
Calibrator.

The power amplifier assembly consists of an RF power amplifier and a 500-volt power supply.

MEASUREMENT

Table IX-II, which shows a list of the measurements, by channels, that were made on a typical missile, is included in this section. This list is used as an example of the type of measurements made in the missile during flight. The measurements allocated to each channel can be changed from missile to missile.

Although more than one measurement may appear for one channel, it does not necessarily mean that the channel is commutated. More than one measurement can be made per channel when one measured quantity no longer exists at the time the next quantity to be measured occurs. Another situation that allows for multiple measuring by a channel is when one measurement is more or less constant and another changes radically. A graphic presentation of this might show a gradually changing base line, representing one measurement, and superimposed on it, an occasional pip representing another measurement.

Table IX-I — OSCILLATOR DEVIATION CHART

Channel	Lower 7.5 % (5 volts)	Center	Upper 7.5% (0 volts)
1	370	400	430 (Not used on REDSTONE)
2	518	560	602
3	675	730	785
4	888	960	1,032
5	1,202	1,300	1,396
6	1,572	1,700	1,828
7	2,127	2,300	2,473
8	2,775	3,000	3,225
9	3,607	3,900	4,193
10	4,995	5,400	5,805
11	6,799	7,350	7,901
12	9,712	10,500	11,288
13	13,412	14,500	15,588
14	20,350	22,000	23,650
15 (commutated)	28,328	30,000	32,456
16	37,000	40,000	43,000
17	48,562	52,500	56,438
18	64,750	70,000	75,250

Table IX-II – MEASURING PROGRAM
(Typical Missile)

Channel	Frequency (cps)	Measurement	Measurement Range
2	560	H_2O_2 Flow Rate	0 to 7 lb/sec
3	730	(a.) Alcohol Flow Rate (b.) Bursting Diaphragm-Cylinder Fin IV	0 to 25 gal/sec
4	960	(a.) LOX Flow Rate (b.) Bursting Diaphragm-Cylinder Fin II	0 to 25 gal/sec
5	1,300		
6	1,700		
7	2,300	(a.) Input to Step Switch (b.) Tilting Program (ST-80)	0 to 180°
8	3,000	(a.) Input to Step Switch (b.) Tilting Program (ST-80)	0 to 180°
9	3,900	Inverter Frequency (1800 VA)	400 ± 0.25 cps
10	5,400	(a.) Deceleration Switch (b.) Turbine rpm	0 to 5000 rpm
11	7,350	Pressure Combustion Chamber	0-400 PSIA
12	10,500	(a.) Acceleration Longitudinal (b.) Acceleration Longitudinal (c.) Acceleration Longitudinal	+1 to +6g +0.5 to -1g 0 to -6g
13	14,500	Lateral Displacement, Fine	500 m/rev
14	22,000	Range Displacement, Fine	1 km/rev
15	30,000	See Table 3	
16	40,000		
17	52,500	(a.) Range Velocity, Fine (b.) Cutoff Computer Output	10 m/sec/rev
18	70,000	(a.) Lateral Velocity, Fine (b.) Take-off (c.) Cutoff	10 m/sec/rev

Table IX-III – MEASURING PROGRAM
Channel 15 Inputs
(Typical Missile)

Number	Measurement
1	(a.) Emergency Cutoff (b.) Pressure in Air Bottles
2	Pressure in Explosion Cylinders
3	Instrument Compartment Pressure
4	(a.) Air Vane 1 Deflection
5	(a.) Air Vane 2 Deflection (b.) Jet Vane 2 Deflection
6	(a.) Air Vane 3 Deflection (b.) Jet Vane 3 Deflection
7	(a.) Air Vane 4 Deflection (b.) Jet Vane 4 Deflection
8	Air Bearing Low-Pressure Supply
9	Top Jets Air Pressure
10	Platform Pitch Position – Minus Program
11	Platform Yaw Position
12	Platform Roll Position
13	(a.) Explosive Screw 1 (b.) Explosive Screw 2
14	Angular Velocity Pitch
15	Angular Velocity Yaw
16	Angular Velocity Roll
17	Pitch Acceleration
18	Yaw Acceleration
19	Voltage Servo Battery
20	Instrument Compartment Temperature
21	Temperature Inlet Air for Air Bearings
22	(a.) Explosive Screw 3 (b.) Explosive Screw 4

Table IX-III – MEASURING PROGRAM (Continued)
Channel 15 Inputs
(Typical Missile)

Number	Measurement
23	(a.) Explosive Screw 5 (b.) Explosive Screw 6
24	Lateral Displacement, Coarse
25	Range Velocity, Coarse
26	Lateral Velocity, Coarse
27	Range Displacement, Coarse

PROJECT MERCURY

FAMILIARIZATION MANUAL

Manned Satellite Capsule

Periscope Film LLC

NASA PROJECT GEMINI

FAMILIARIZATION MANUAL
Manned Satellite Capsule

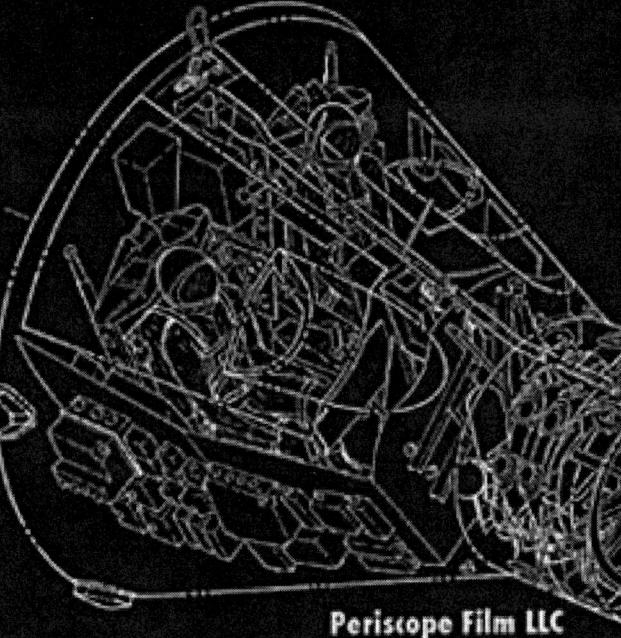

Periscope Film LLC

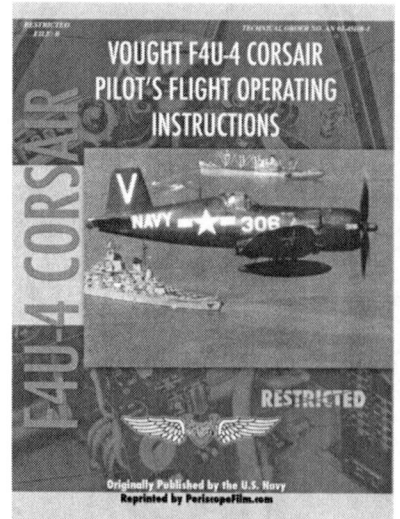

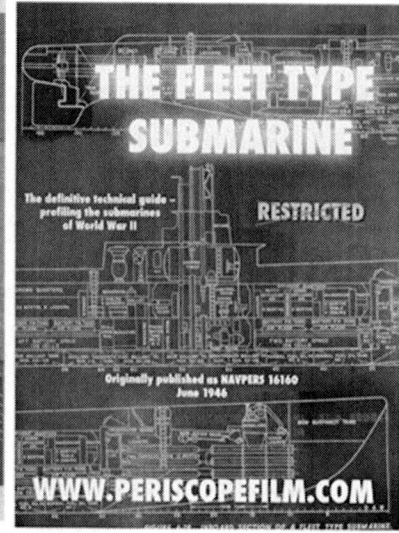

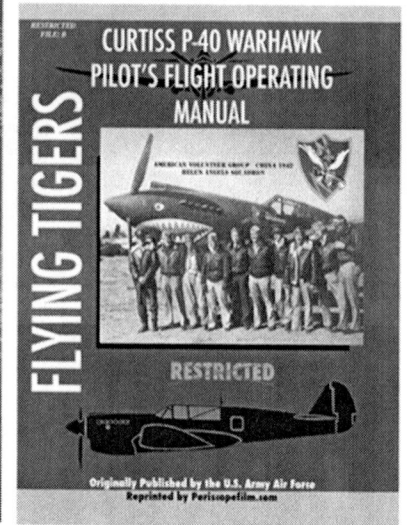

ALSO NOW AVAILABLE
FROM PERISCOPEFILM.COM

©2012 PERISCOPE FILM LLC
ALL RIGHTS RESERVED
ISBN #978-1-937684-80-8
WWW.PERISCOPEFILM.COM

MMS SUBCOURSE NUMBER 151 *EDITION CODE 3*

NIKE MISSILE
and Test Equipment

NIKE HERCULES

DECLASSIFIED

by U.S. Army Missile and Munitions Center and School
Periscope Film LLC

CPSIA information can be obtained at www.ICGtesting.com
Printed in the USA
BVOW050219270313

316566BV00001B/6/P